The Mill on the Boot

The Story of the St. Paul & Tacoma Lumber Company

MURRAY MORGAN

Published in cooperation with the Washington State Historical Society

University of Washington Press   Seattle & London

Dedicated to the memory of
George Corydon Wagner, Jr.,
who wanted the story told

Library of Congress Cataloging in Publication Data

Morgan, Murray Cromwell, 1916–
    The mill on the Boot.

    Includes index.
    1. St. Paul & Tacoma Lumber Company—History.
I. Washington State Historical Society.   II. Title.
HD9759.S72M67   1982   338.7'674'00978   82-16107
ISBN 0-295-95949-5

Design and production: Audrey Meyer
Printing and binding: Vail-Ballou Press, Inc.

Fallers on springboards notch a Douglas fir before they start to
saw (Thomas Rutter, photographer, 1888; Tacoma Public
Library)

Preface

WHEN I was growing up in Tacoma, the St. Paul & Tacoma Lumber Company complex on the tideflats was as much a part of local geography as Mount Rainier. We usually called Rainier "The Mountain," and always called the mill "St. Paul."

Outsiders have come to think of Tacoma in terms of Weyerhaeuser, but though the Weyerhaeuser Timber Company was incorporated in Tacoma and has maintained its headquarters in or near the city, it never cut lumber there. St. Paul did. It had the largest sawmill payroll; dreams centered on the fill land created from a tideflats island known as "The Boot." "Dad's working at St. Paul." "If I can catch on at St. Paul." "I wish we could but the mill's on short shift." "They're hiring again at St. Paul."

When the Washington State Historical Society told me funds had been raised to sponsor a history of the company and asked if I would write it, I accepted with enthusiasm. It was agreed that I would not write an official history, or an economic history. The task I set myself was to write of the role of the company and its leaders in the life of the community. Final decisions on the manuscript would rest entirely with me.

Research has given me the opportunity to look on familiar territory from a fresh point of view. How did the company's original purchase of 80,000 acres of timberland from the Northern Pacific Railroad's land grant influence the development of Puget Sound? What role did St. Paul play in seeing Tacoma through the Panic of '93? How did the company influence politics, and how did politics influence the mill?

What I learned often surprised and sometimes delighted me. In spite of regrets at the disappearance of the virgin forest, I find pleasure in remembering that mines around the Pacific rim are shored up with timbers cut on the tideflats, that the first railroads in China and Korea ran on Douglas fir ties, that wharves in Japan and Liverpool stand on pilings creosoted in Tacoma, and that houses by the million in the States and overseas are created from the processed forests that put to sea on lumber schooners or rolled eastward on the rails.

I especially enjoyed making the acquaintance of the four middle-aged men who nearly a century ago gave up established and honored roles in the Mid west to bring state-of-the-art logging and milling practices to Puget Sound. The names of Henry Hewit, Chauncey Griggs, Addison Foster, and C. H. Jones had been familiar to me, but the people who bore the names were like pictures on a school-house wall. Now they are real people, sometimes quirky but always interesting, part of the continuum that is Puget Sound to me.

In the course of my research I enjoyed the cooperation and hospitality of many of the descendants of the founding fathers. The late Corydon Wagner, Jr., grandson of the founders, conceived the idea of a book about St. Paul & Tacoma Lumber Company after its merger into St. Regis. John and Will and Henry Hewitt shared his interest, as did Chauncey Leavenworth Griggs, Howard and Tom Meadowcroft, and others with familial connections to the company. The late Mrs. Mabel Kimberly Gilbert of Neenah, Wisconsin, was especially helpful in my search for material on the Hewitts, and Mrs. William Pettit of Lexington, Kentucky, on the Fosters.

Bruce Le Roy, director of the Washington State Historical Society, and the late E. K. Murray, attorney for the society, worked out many details on the project.

As always I am indebted to many librarians and archivists: Morris Skagen, Richard Aiken, and Lorraine Hildegrand of the Tacoma Community College library; Frank Green and Jeanne Engerman of the Washington State Historical Society; Gary Reese, Jean Gillmer, and Brian Kamens of the Tacoma Public Library; Robert Monroe and Dennis Andersen of the Special Collections of the University of Washington Library; Richard Berner and Kathryn Wynn of the University of Washington Archives; Nancy Pryor of the Northwest Collection of the Washington State Library; Florence Fry, librarian for the *Seattle Post-Intelligencer;* and Laura Clark, photo librarian for the *Tacoma News-Tribune.*

Also supplying information were the staffs of the Pacific Northwest Collection, University of Washington Library, the Manitoba Provincial Archives, Winnipeg Library, Fargo (N.D.) Public Library, University of North Dakota Library, Menasha (Wisc.) Public Library, Neenah (Wisc.) Public Library, Appleton (Wisc.) Historical Society, Oshkosh (Wisc.) Public Library, University of Wisconsin Library, University of Chicago Library, Everett (Wash.) Public Library, Hoquiam (Wash.) Public Library, Forest History Association, Tacoma Fire Department, U.S. Corps of Engineers, the U.S. Coast and Geodetic Survey, and especially the Minnesota Historical Society.

My wife Rosa was indispensable in every phase: research, writing, editing, proofreading, indexing, hand-holding.

Murray Morgan
Trout Lake, August 1982

Contents

Illustrations

The Mill on the Boot

THE FIVE TRACKS.

The switchbacks on the Cascade Division carried Northern Pacific trains across Stampede Pass before the tunnel was completed through the shoulder of Mount Rainier (from The West Shore, 1887)

June 4, 1888

THE WEATHER had changed. The unseasonal rain had lifted and the sky was clear as the little train—locomotive, baggage car, wooden-seated passenger car, and a parlor car called "The Glacier"—labored up the eastern slope of the Cascades.[1] The grade was steep on the temporary track the Northern Pacific had laid over Stampede Pass, and the locomotive panted heavily as it sashayed back and forth up the switchback. (Hundreds of feet beneath the train, deep in the mountain, workmen panted too as they chipped away at the face of the tunnel that was to make economical the movement of heavy freight between Puget Sound and the Mississippi River.) Crossing the crest of the range, the train began a cautious descent, the engineer spilling sand ahead of the wheels, the brakemen alert.

The men in the parlor car studied the landscape with new interest. They were in Washington's Pierce County now, the area they intended to develop. Leaving Yakima the tracks had climbed through pine; descending toward Tacoma they passed first through scattered clusters of spire-like alpine firs, their deep green foliage conspicuous against the glazed snow of spring, then through stands of noble firs with blue green needles and gray bark divided into irregular plates. Farther down the slope, they entered true forest. There was some hemlock and a scattering of cedar, but the basic tree was Douglas fir. The forest was dense, many of the trees immense—150 feet or more in height, the cinnamon brown boles 5 to 8 feet thick at the butt. Mile after mile the train tunneled through deep shade, the stench of coal smoke faint in the fresh scent of evergreen, the very whistle of the locomotive hushed.

The men in the parlor car were all lumbermen. They had followed the receding forests from New England to the Great Lakes and had made fortunes in timber. One of them owned more pine trees than any capitalist who ever lived. They thought comfortably in multiples of 100,000 board feet, but they were awed by this forest they had come to buy, and fell, and cut, and ship around the world as building stuff for homes and industries.

They came off the mountain into the foothills. The train passed through a series of short tunnels and over little bridges spanning the Green River. Broken patches of forest alternated with truck farms and small fields of wheat. The tracks left the Green and curved away down a last steep slope into the valley of the Puyallup. The blunted cone of Mount Rainier, its slopes still shining with winter snow, loomed above the hills to the southeast. The vine-covered trellises and ventilator-topped drying barns of hop farms filled the rich bottomland along the right-of-way. At the one-street village of Orting, the train paused to discharge from the passenger coach a party of booted, mackinawed men carrying canvas bags: timber cruisers brought west from Wisconsin and Michigan to examine the 80,000 acres of forest that the men in the parlor car were about to acquire from the Northern Pacific Railroad Company in the largest purchase of timberland in the nineteenth century.

Beyond Orting the NP tracks paralleled the Puyallup until they emerged at the head of Commencement Bay and crossed the boggy tideflats on a long trestle. The train shuddered to a stop alongside a wooden shed that served as a station near the eastern end of Pacific Avenue in the raw boom town of Tacoma.

The visitors disembarked onto a plank platform and looked up the avenue, a dirt street lumped with the remains of stumps and flanked by wooden sidewalks, wooden storefronts, and a few squat blocks of masonry that were the pride of downtown. After St. Paul, with its population of 81,000 and its gracious Summit Avenue on which stood the Midwest's most expensive mansions, this was hicksville: population 15,000, according to a generous Chamber of Commerce estimate. But the midwesterners liked what they

Loggers from the Midwest found themselves operating in a new dimension of forest (from West Coast Lumberman)

saw. In the four months since their first visit to Tacoma the town had grown, and with their help they could foresee no end to its growth. They were not thinking in terms of a greater St. Paul; what they had in mind was a greater New York.

They were not given to pipe dreams, these parlor-car pioneers of 1888 who looked at mud streets and fire-trap stores and envisioned the metropolis of the Pacific. They were men of substance and personal accomplishment. Colonel Chauncey Griggs and his partner, Addison Foster, had prospered in St. Paul real estate, Wisconsin lumber, Minnesota fuel, and Dakota wheatlands; Griggs's twenty-seven-year-old son Herbert was a graduate of the Yale Law School. Henry Hewitt, Jr., of Menasha, Wisconsin, was said by some to be the richest man in his home state, or at least in Winnebago county, and said by himself to be the world's premier accumulator of pine stumpage. Charles Hebard Jones, his partner and brother-in-law, owned a lumber company in Menominee, Michigan, and was associated with Hewitt in other ventures. P. D. Norton, another of Hewitt's associates, had operated sawmills, a paper mill, a flour mill, a bank, and a smelter. Peter J. Salscheider of Green Bay was reputed to be the best millwright in Wisconsin. They were entitled to dream big dreams.

They planned to incorporate a company, to build the biggest sawmill in the world, to invade the midwestern market and battle the pine mills of the upper Mississippi on their home grounds. Nor was that all. They would build a rail-road, not just a logging line but a road up to transcontinental standards. They would contract to construct other lines for the Northern Pacific. They would mine coal, build wharves, open a general store, found a bank, while running the businesses that had already made them rich and keeping their eyes open for other opportunities.

The party was met at the station by another man of big ideas, George Browne, vice-president of the Tacoma and Fern Hill Street Railway Company, speculator in Tacoma real estate, nephew of the vice-president of the Northern Pacific. He took them by carriage to the two-story wooden courthouse on C Street (today's Broadway) where County Auditor Edward Huggins, a gnarled relic of the Hudson's Bay Company period of proprietorship in Pierce County, accepted and filed articles of incorporation for the St. Paul & Tacoma Lumber Company, with capital stock of $1,500,000 divided into 15,000 shares.

Incorporation accomplished, the party proceeded to the Tacoma, a handsome hostelry, four years old, perched at the very edge of the downtown bluff. There they dined in a private room overlooking the bay. Below them, on a mudflat island known as "The Boot," was the site of their proposed mill. To the south and east, toward the mountain glowing pink in the sunset, stretched their forest empire; to the north and west, the waters of Puget Sound opened onto the markets of the world.

After a leisurely dinner and the best wine and brandy in the house, the party strolled down Tenth Street to 936 Pa-

Pacific Avenue, Tacoma, 1887 (pen and ink sketch by J. T. Pickett; from Northwest Magazine, *June 1887)*

cific Avenue, where George Browne made available the office of the Tacoma and Fern Hill. In a room lit by one of the town's few electric lights they organized the St. Paul & Tacoma Lumber Company. Colonel Griggs was elected president; Foster, vice-president; Hewitt, treasurer; and Browne, secretary. Jones was put in charge of all lumber and mining operations, Salscheider was appointed millwright, and Norton was assigned accounting duties. Herbert Griggs, it was agreed, would open a private law office and handle most of the company's legal affairs. That done, they called in the press. They did not ask for reporters, they sent for editors.

"The first thing to be done, and at once," said Colonel Griggs, speaking for the company, "will be the construction of a branch of railroad from the main line of the Northern Pacific across the Puyallup to the mill site on The Boot on the flats below the city." Continuing:

Mr. Salscheider will begin the work of constructing the first of two mills immediately. It will be erected on The Boot below the city and will be completed and ready for cutting lumber by December first. The mill will have a capacity of 60,000,000 feet of lumber annually. As soon as it is completed another mill will be commenced with a much larger capacity and pushed to the earliest possible completion.

Within thirty days the construction of the Tacoma Southern railway will be commenced, leading to our timberlands, and will be built with all possible speed.

We are all delighted with Tacoma, and in a very few days Mr. Hewitt and myself will begin the erection of our residences, as we propose to make this city our future home.

The next day's headlines read:

A GREAT ENTERPRISE

THE MONSTER MILLING COMPANY
OF TACOMA ORGANIZED
—*Tacoma Ledger*

A BONA FIDE BOOM

THE ST. PAUL & TACOMA LUMBER CO.
—*Tacoma News*

Founding Fathers

FOUR MEN dominated the formation of the St. Paul & Tacoma Lumber Company. They came in matched pairs: Chauncey Griggs and Addison Foster from St. Paul, Henry Hewitt and Charles Jones from Menasha and Menominee. Though their interests blended and the four were associated with St. Paul & Tacoma throughout their lives after 1888, their early ties remained strongest. It was always Griggs and Foster, Hewitt and Jones.

Chauncey Wright Griggs

Chauncey Griggs, the eldest of the founding fathers, was born in Connecticut on the last day of 1832 to a family that had roots two hundred years deep in American history.[1] He attended public school in Tolland, Connecticut, dropped out briefly to work as a clerk in a country store, but returned and was graduated from Monson Academy at the age of eighteen. Almost immediately he was appointed headmaster at a private school, a title made less impressive by the fact that there was but one other employee. When he was twenty, Chauncey received fifteen hundred dollars from his father. So did a brother. They pooled their capital and opened a general store in Willimantic, Connecticut. When the first year netted only a hundred dollars profit, Chauncey sold out to his brother and went west.

There followed a period of wandering. He took a course in bookkeeping in Detroit and caught on with a small bank, which paid an infinitesimal salary. This hardly beat storekeeping. He moved to Ohio and bought an interest in a livery stable; failing to clean up in the horse business, he traded his share for a small store in Kent, Ohio, only to find his horizons circumscribed by counter and cashbox. He sold the store, invested in drygoods, two teams of horses, and a wagon, and set out as a peddler. He reached the end of that line in Montezuma, the county seat of Poweshiek, in southern Iowa, where he disposed of wagon and teams for enough to get back to Detroit. His brother George had entered the wholesale furniture business there, and Chauncey joined him. After two more years of fraternal enterprise he took to the road again, heading west, this time to St. Paul, Minnesota, where he struck root.

St. Paul, a town of about ten thousand in 1856, stood at the head of navigation on the Mississippi. The jumping-off place for the upper Midwest, it was a magnet for young men of ambition. "Every man here, to use a western expression, 'is a *steamboat*,' " a woman traveler wrote that year. "It is a strange medley indeed, that which you meet aboard a Mississippi steamer. An Australian gold-hunter, a professor in an eastern University going out to invest in Minnesota, a South Carolina boy, with one thousand dollars and a knowledge of double-entry. . . ."[2]

Chauncey opened a grocery store on the corner of Third Street, then a general store at Seven Corners in downtown St. Paul. He dealt in real estate. He bought an interest in a small lumber mill on the west side of town. He won a contract to grade Bench Street. By 1859 he had accumulated five thousand dollars and was doing well enough to return briefly to Connecticut to marry in the Ledyard Congregational Church, Martha Ann Gallup, whom he had met earlier while she was visiting in Detroit.

The Griggses were a popular couple in fast-growth St. Paul, a young person's town. Both loved to dance, favoring reels, quadrilles, and quicksteps. A contemporary remembered Martha Ann as "an elegant woman of superior ability, of remarkable energy and charming manners, a young-old settler so to speak, whose influence was always for good." Chauncey was "a fine-looking man, well-proportioned and possessing excellent business qualities. He is quiet in his movements, cool and deliberate, but effective. He is exceed-

ing pleasant and social in his nature, yet back of all this he is shrewd and scheming."

Griggs's business affairs prospered, and the family flourished. The first child, Chauncey Milton (he went by the Milton), was born early in 1860; the second, Herbert (Bert) Stanton, a year later. Then came the Civil War.

The war posed a problem for a patriotic man of twenty-eight years with a young wife and two children under the age of two. It was Martha Ann who persuaded Chauncey she could look out for herself. She took the children back to the Gallup farm in old Ledyard, "staging it," as her son Milton recalled. "She laid up to wash baby clothes, she took the steamboat a way, the stage another piece, finally the railway from some place in Wisconsin to Milwaukee; then the boat across the Lake to Grand Rapids, Michigan, then rail to Detroit, where she was among other Griggs, who kept her there a month resting up before she went on East by train back to old Connecticut, to the old Gallup farm."

As for Chauncey, he sold most of his business interests and helped recruit members for a new regiment, the Third Minnesota Voluntary Infantry in which he enlisted as a private on October 12, 1861. Three days later he was elected captain of Company B. After brief training, the Third Minnesota was assigned to guard supplies being moved through Kentucky and Tennessee. On May 1, 1862, Griggs won a field promotion to major for gallantry during a skirmish. Four weeks later he was commissioned lieutenant colonel.

In July the Third Minnesota took part in its first real

Chauncey Griggs as a member of the Minnesota Senate (*from* Northwest Life, *Jan. 1944; Minnesota Historical Society*)

Martha Ann Griggs, ca. 1865 (*from* Northwest Life, *Jan. 1944;
Minnesota Historical Society*)

battle, and was humiliated. A Confederate cavalry force under Colonel Nathan Bedford Forrest, assigned to delay the Union drive on Chattanooga, staged a surprise attack on the Union supply depot at Murfreesboro, Tennessee.

The Confederates struck the Ninth Michigan and supporting units of General T. L. Crittenden's provost guards. The Third Minnesota was on the Stones River about a mile and a half from the center of town. When they heard the southern guns, the regiment marched in a flank movement to an open plain looking toward the town and halted. Two pieces of artillery were placed on the right and left, and two companies of skirmishers were deployed in front of the lines.

"After a halt of two hours," Griggs wrote in his review, "the regiment was advanced a short distance. About eight o'clock the entire force in front of our line charged and were handsomely repulsed with a heavy loss in killed or wounded, while we sustained the wounding of one man."

While the Confederates were being repulsed in this attack, a part of their force swung to the left and struck at the camp the Third Minnesota had vacated in the morning. It was defended by a guard of twenty men, the soldiers on the sick list and the teamsters. They fought desperately, holding off the attack and waiting reinforcement. Griggs asked permission of his superior officer, Colonel W. C. Lester, to lead a company to relieve them, Lester refused to divide his main force. The base camp was overrun.

Shortly afterwards the commander of the Ninth Michigan called on the Third Minnesota for help. Griggs urged that it be sent. "There is no doubt we should have punished Forrest severely and probably captured his entire command," he said in his review. But again Colonel Lester refused to commit his troops. The Ninth Michigan was overrun and the Third Minnesota was left alone facing Forrest's forces. The Confederate commander asked for a conference. Colonel Lester went out to meet him under a flag of truce and returned from the talk ready to surrender.

"I stated that I could see no good reason for it," Griggs recalled, "but on the contrary expressed the opinion that we could hold our position until re-enforcements reached us from Nashville, and even if aid did not come, I had such unbounded confidence in the charge of our men that I believed if we were led against the enemy we could whip him in fair and open context. . . . Lester then called all of the Company commanders to a council. Here the matter was freely discussed, a viva voce vote was taken and nearly all voted against the surrender. Subsequently another conference was called and officers were required to vote by ballot. I called upon all who voted for fighting to do so with open ballots, so we should be rightly reported. A majority being on the side of the commander the most ignominious surrender was made."[3]

The Confederates paroled the Union enlisted men and allowed them to return home. The captured officers were marched to Madison, Georgia, and imprisoned in a tobacco warehouse, where the other inmates greeted them cheerily with cries of "fresh fish, fresh fish." One of Griggs's fellow officers, Captain Everett Foster (Addison Foster's brother), wrote of their captivity:

Roll call in the morning at about six o'clock, breakfast usually consisted of fresh bread and sometimes wormy bacon, the bread was baked in large cakes perhaps two inches thick and twelve inches in diameter, baked hard, similar to hardtack, and all prepared at an outside camp and brought in by a detail from the civil prisoners. The food for breakfast, dinner and supper was about the same, except that we were occasionally served soup and coffee made from parched rye.

After only three months' confinement, Griggs was released in an exchange of prisoners. He returned to St. Paul in October. The Third Minnesota was called back into service on December 1, 1862, less than five months after its surrender. Officers who had voted to give up were dismissed. Griggs was promoted to full colonel and given command.

The regiment was assigned to protect a five-county area in Tennessee. During the winter Griggs's men captured more than a thousand Confederate soldiers and irregulars as well as $5,000,000 worth of cotton and salt, and destroyed two factories producing cloth for the southern forces. Griggs recalled, however, that "this irregular warfare was not pleasant to me nor to the officers and men under me, and I applied to have the regiment sent to the front."

Request granted. The Third Minnesota was transferred to the Mississippi where it faced General Joe Johnston's troops at Vicksburg. During the siege, Colonel Griggs came down with malaria. Desperately ill, he refused hospitalization until after the city was surrendered to General Grant on July 4, 1863, opening the entire Mississippi to Union transportation. Then on the advice of the regimental surgeon, Griggs was invalided out of the army.

Returning to St. Paul fifty pounds underweight and still plagued by intermittent chills and fever, Griggs did not feel up to big city life. He sent for Martha Ann and the boys and moved to the backwater town of Chaska, thirty miles to the west, where he gradually eased back into business. He dealt in real estate. He manufactured some bricks. He sold wood and coal. He bought a mill. He stood for the state legislature and though still a Democrat in a Republican era he was elected, first to the house, then to the state senate. Four years after the war he again felt ready to challenge the world. St. Paul called.

On August 20, 1869, Chauncey Griggs entered into partnership with an old acquaintance, James Jerome Hill, a thirty-year-old Canadian who had arrived in St. Paul the same year Griggs did, and who had made a name for himself as a freight-forwarding specialist. They planned to deal in coal and wood.[4]

At the time when Griggs and Hill became partners, changes in technology, economics, and politics were creating new opportunities in the valleys of the upper Mississippi and the Red River of the North. The headwaters of the two river systems are separated by low ridges and almost imperceptible tilts of land. During spring floods it is sometimes possible to canoe between them. Rain from a single cloud can drain into the Mississippi and reach the Gulf of Mexico, and into the Red and flow north to Hudson's Bay.

Even in the days when St. Paul gloried in the name of Pig's Eye, its merchants already serving the upper Mississippi dreamed of adding the Red to their trade area. Though the northward flowing river was tantalizingly near, there were problems in doing business with its inhabitants. Roads were nonexistent, trails few. The international border bisected the river. The Hudson's Bay Company controlled the British portion of the Red.

The north-of-the-border merchandise that reached St. Paul—mostly beaver pelts and buffalo robes—was brought south in two-wheeled carts constructed entirely without metal by the métis (French–Indian "half-breeds"), who annually formed a caravan of creaking wagons for a round trip between Winnipeg and the Mississippi. The Red River wagons were colorful but inefficient. Merchants wanted a better connection.

In 1858, only two years after Griggs and Hill made their separate arrivals in St. Paul, the business community offered $1,000 to the first man to put a steamboat on the Red River. Anson Northup, an experienced Mississippi riverboat man, said he would try if the prize were doubled. He was told to go ahead. Northup ran his little sternwheeler *North Star* up the Mississippi to Crow Wing, had her disassem-

bled, and hauled the machinery, cabin furniture, and lumber in thirty-four ox-drawn carts for 130 miles along the Woods Trail to the upper Red. Reassembled and reborn as the *Anson Northup,* captained by Anson Northup in person, the sternwheeler thrashed downstream to Winnipeg, to be greeted by a cannon salute, the chiming of church bells— but no return cargo.

The $2,000 prize did not pay Northup's costs in putting his namesake in service, nor did available freight and passengers make it profitable to keep her running. He sold the paddlewheeler to another St. Paul entrepreneur, J. C. Burbank, for $8,000. After running the vessel for a season, Burbank decided she was "a lumbering old pine basket, which you have to handle as gingerly as a hamper of eggs." He tied her up at a dock near the border where she stayed until she too gave up, rolled over and quietly sank. The Hudson's Bay Company salvaged her engine and set it to sawing wood.

Burbank tried again with a bigger boat. His *International* was built at Georgetown on the American side of the boundary. Powered by the engine from a sternwheeler whose captain tried to run between the Mississippi and the Red during the spring flood only to be stranded midway, the *International* was 136 feet long, 26 feet wide, which proved to be too big for a river with so many turns that its admirers claimed, "when straightened out the Red will make a canal from Winnipeg to Mexico City." The *International* had to be laid up during the winter freeze, summer drought, and

The sternwheeler International *at Fort Gary, north of Winnipeg on the Red River of the North* (*Manitoba Archives*)

whenever Indian demands for hush money (they protested that the steamer's noise drove away fish, game, and the spirits of their ancestors) became exorbitant.

The Hudson's Bay Company was an even greater problem. "They did not want immigration and trade, nor mails or other appearances of civilization," Burbank claimed. "The country could not be opened up against the interest of that powerful organization." So he sold the *International* to the HBC which used it for hauling company goods but did not offer regular freight service. The HBC lost its power to control events in the Red River Valley when Upper Canada, Lower Canada, Nova Scotia, and New Brunswick were united as the Dominion of Canada in 1867 and the Province of Manitoba was organized in 1870.

Such was the situation when Griggs and Hill decided to put a steamboat on the river. Since neither was an experienced riverboatman, they recruited Captain Alexander Griggs of St. Paul to build and run a steamer.[5] (Let not his name confuse the story: Alexander Griggs was not related to Chauncey Griggs.)

Captain Griggs reached Fort Abercrombie on the Red River in July of 1870. Finding no lumber available for shipbuilding, he took a gang of workers upstream to Otter Tail Lake, where they cut oak and pine, rafted the logs to a mill for sawing, built seven barges with some of the lumber, and floated downstream to McCauleyville just before the river froze. Twenty-five workmen, a few of them skilled, labored through the bitter Dakota winter and had a sternwheeler ready to hit the water by April 12.

The *Selkirk,* as Griggs and Hill named her in tribute to the Scottish lord who created the Red River Colony, was twin-stacked, blunt-nosed, and flat-bottomed, designed to float on spit and strong wishes. She was painted white and laced with more gingerbread trim than a cuckoo clock. It took a crew of fifteen to operate her—Captain Griggs, two pilots, two engineers, two firemen, two clerks, a steward, a cook, and two maids, to say nothing of a dozen roustabouts who went along to handle cargo.

Soon after her launching the *Selkirk* was on her way downstream. Accounts vary as to the maiden trip. One says she was "loaded to her hurricane decks with freight but had only a few passengers." Another story credits her with "175 tons freight and 115 passengers." Yet another puts "the gallant and delighted Jim Hill himself on the hurricane roof." At 110 feet, she was shorter than the *International* and was expected to have less trouble rounding bends, but a passenger reported "we go from one bank to another, crushing and crashing against trees, which grow down to the waterside and in consequence of this curious navigation we never really go on for three minutes at a time." The *Selkirk* made it to Winnipeg and back in twelve days and a newspaper report speculated that Griggs and Hill netted almost enough on the trip to pay the $7,000 cost of putting her on the river.

The sternwheeler Selkirk *loads wheat and farm machinery during a pause at the Moorhead levee in North Dakota, 1877 (Manitoba Archives)*

All through the summer and fall, the *Selkirk* plied the Red, starting either at Georgetown or, when the water was low, at Frog Point at the foot of Goose Rapids. Chauncey Griggs made the trip in July, accompanying a party of politicians, military men, and journalists who were visiting the Red River country at the invitation of Jay Cooke, the promotion-minded financial agent for the Northern Pacific. Griggs's companions included the vice-president of the United States, Schuyler Colfax, traveling somewhat incognito; Senator William Windom of Minnesota; Thomas H. Canfield, vice-president of the Northern Pacific; several generals of the army, most notably William Tecumseh Sherman, Phil Sheridan, and Nelson Miles; and the granddaughter of naturalist James Audubon. Of the journalists, Charles Dana of the *New York Sun,* Samuel Bowles of the *Springfield Republican,* and J. H. Harper of Harper Brothers were the most famous, but the *New York Herald, Chicago Tribune,* and *Baltimore Sun* were also represented. Articles appeared in sixty daily papers. The Red River boom was well launched.

While in Canada, Colonel Griggs paid a visit to the governor of Manitoba. (A Canadian businessman complained that Griggs was given too much time: "It would seem as if fate had decreed that I should not enter the august presence. I waited, not patiently, for ¾ of an hour for Col. Griggs' exit.") The conference almost certainly concerned negotiations then in progress over international steamboat politics.

The Hudson's Bay Company had responded to the Griggs and Hill entry into their territory by putting the *International* into regular service against the *Selkirk.* Hill and Griggs, who had anticipated such a move, immediately protested to the American government that only American vessels could carry freight or passengers on American rivers. The HBC got around that by transferring title of the *International* to Norman Kittson, a company employee who had taken American citizenship. They also ordered construction of a second vessel, the *Dakota.* Hill and Griggs then prodded the Treasury Department to enforce a regulation requiring foreign goods that were transshipped across the United States under bond (as was the case with much HBC goods bound for Winnipeg) to be carried by firms under bond in the United States. Foreigners could not get such bonds so the *Selkirk* was the only ship eligible to carry the freight.

Such manipulation of regulations offered a poor foundation for business. Griggs's conference with the governor of Manitoba seems to have been the start of an attempt to put Red River transportation on a more rational basis. It was followed by a visit to St. Paul by Donald A. Smith, the ranking HBC official in Canada. After talking with Griggs and Hill, Smith wrote guardedly to Beaver House, the company headquarters in London, that he was bringing them "an agreement which I think will be greatly to the advantage of the Company, which has just been entered into between Mr. Kittson and Messrs. Hill, Griggs & Co. of this place for regulating the carrying business between Moorehead or Georgetown and Garry this ensuing season. . . ."

Another and not less important effect will be to prevent the possibility of any such difficulties incurring as those encountered with the Bonding system which were so inconveniently felt the early part of last season."

The deal called for the formation of a new company, the Red River Transportation Company, nominally American but combining the steamship interests of Hill and Griggs with those of the Hudson's Bay Company. Norman Kittson, HBC's American agent, would be manager, supervising the operations of the *Selkirk,* the *International,* and the *Dakota.* Smith cautioned the home office in another letter that the Hudson's Bay Company's "relations in regard of these shares, you are aware, are very peculiar in consequence of the law of the United States which excludes any other than their own citizens from taking part in the navigation of their inland waters. We have therefore to be very careful not to compromise the company in any way."

The new company, enjoying a de facto monopoly on the river, decided the minimum freight rate would be $1.50 per hundred pounds, except for Hudson's Bay Company goods, which would receive a 50 percent discount. In practice, prices never fell close to the minimum. During the first season the Red River Transportation Company charged a basic $3.50 per hundred, which resulted, according to Smith's report to Beaver House, "in a very handsome profit, highly satisfactory."

The next year was even better. The company acquired two more vessels, the *Cheyenne* and the *Alpha,* both built for competitors who went broke before getting on the river. In 1874 passenger traffic doubled and freight totalled 16,300 tons. Rates were lowered to $3 per hundred, but Manitoba merchants felt they were being unconscionably squeezed. They raised $50,000 to finance a new line, the Merchants International Steamboat Company, to compete with Griggs, Hill, and HBC.

Merchants International had two steamboats assembled at Moorehead. The *Manitoba* and *Minnesota,* each 135 feet long and 31½ feet wide, took to the water in the spring. They were well equipped and the *Minnesota* sported such unusual amenities as a tame cinnamon bear and a pig named Dick that engaged in wrestling matches on deck. But the management, inexperienced and overly optimistic, set rates so low that even with full loads each run lost money. Then the *Manitoba* was rammed and sunk by the *International* in an accident which may not have been accidental. She was raised and put back into service at considerable expense. Then fire damaged the *Minnesota.* Merchants International tossed in the towel. The Griggs–Hill–HBC combine faced no serious competition until the completion of the railroad between St. Paul and Winnipeg in 1878 ended the golden period of steamboating on the Red River of the North.

The steamboat venture, though exciting and profitable, was a sideline for Griggs and Hill. What had brought them together in 1869 was their interest in fuel. Griggs had dealt successfully in firewood, Hill in coal.

The International *carried a contingent of Menonite settlers to Winnipeg in 1874 (Manitoba Archives)*

19 *Founding Fathers*

Soon after forming their partnership, they brought John A. Armstrong of Minneapolis into the firm. He had distribution connections and, more important, a concession for taking hardwood logs from public lands in northern Minnesota. William A. Newcomb, another experienced fuel dealer, joined them a few months later. The four partners split profits into five shares, one for each man, the fifth to be used by Griggs and Armstrong "for the benefit of the whole concern."

Wood was the standard fuel in heavily timbered Minnesota, but coal was gaining importance as the expansion of railroads and the growth of the river steamer fleet made possible the movement of bulk cargoes. Hill, who had worked as a freight agent, recognized that coal would become increasingly the source of industrial energy, a view to which he converted Griggs.

Coal was an unfamiliar commodity to consumers in Minnesota, confusing in its variety of sizes, grades, and chemical properties, and unduly expensive because dealers imported small quantities directly from Pennsylvania and accepted as inevitable the high shipping costs. Hill, Griggs, and Company sought to establish grades, educate users, and get better freight rates by buying large quantities of Pennsylvania anthracite at Chicago prices, then arranging bulk shipments to St. Paul.

Hill, Griggs, and Company joined other dealers to form the Minnesota Coal Association. Its mission was to maintain price levels through informal agreement, and to present a united front to the railroads on rate questions. United the dealers were not; the association disintegrated within a year—but demand for coal expanded anyway. Griggs and Hill sold 35,000 tons of all types of coal in their first year of operation. By 1873 they had sold 5,000 tons of anthracite alone, and that in the face of the Panic of 1873, which touched off the nation's first industrial depression.

Although coal consumption increased even in bad times, the Panic strained the Griggs–Hill relationship. Griggs was bred to the New England tradition of community responsibility; Hill, only eight years Griggs's junior, was of a later and less benevolent type. For him, the enterprise was all. What was best for Hill, Griggs, and Company would, in the long run, be best for the community. His responsibility was to be successful.

So it was to Colonel Griggs that people wrote asking for indulgence. Supplicants ranged from the mistress of an academy for young ladies who begged extra time on fuel bills because eight of her pupils simply had not come back after Christmas vacation, to the president of the St. Paul and Pacific Railroad, who asked a loan of a thousand dollars to fend off a powerful creditor. Griggs looked at his customers' personal problems; Hill looked at the bottom line. Hill took the older man to task for his generosity and optimism. "Spent forenoon talking over matters with C. W. Griggs," he noted in his diary in May of 1873 after Griggs quoted prices to a customer six months ahead of promised delivery—a risky act in a commodity subject to unpredictable fluctuations. "He expressed much sorrow for his conduct . . . and desires to try and do better."[6]

The relationship between the hard-driving, one-eyed Canadian and the urbane New Englander was more profitable than pleasurable. On May 1, 1875, Griggs sold out his interest in the company to Hill for $35,000. They parted in cool friendship, knowing they would soon test their theories of business conduct in competition with each other for the Twin Cities coal trade.

Hill immediately formed a new company—Hill, Acker, and Saunders—and won an agreement from the Anthracite Association to handle all its sales northwest of Chicago. Griggs formed a partnership with General R. W. Johnson, another Civil War veteran, and sought to purchase anthracite from hard coal miners who did not belong to the Anthracite Association. The depression led to price-cutting, and Griggs found coal that he was able to offer at prices below those Hill was asking.

Hill proposed to allow Griggs the small lot market, sales of less than 300 tons, but the Anthracite Association demanded an all-out price war. "You must adapt your methods to theirs," the secretary of the Association declared in a letter to Hill, "and not be scared by any price however low but go down to theirs with good grace."[7]

Battle was joined. Whenever Griggs lowered prices on anthracite, Hill matched them, although as the larger dealer, selling far more coal than Griggs, he suffered greater losses. Hill had support from the railroads, however, which gave him rebates under the table, and from the Anthracite Association, which made up part of his losses.

The contest between the former partners ended in compromise. A year to the day after they had dissolved their partnership, Griggs and Hill joined the other major distributors in the Twin Cities area in an arrangement to divide the market. Hill and Acker were to sell 39 percent of the Twin Cities' anthracite; Saunders, 24 percent; the St. Paul Coal Company, 18.5 percent; and Griggs and Johnson, 18.5 percent apiece.

After a year of unofficial price-fixing and market-sharing (then quite legal), the companies merged into a corporation—the Northwestern Fuel Company, Inc., with Hill as president and Griggs as vice-president—which controlled anthracite distribution on the upper Mississippi for many years.[8]

Associates again, their rivalry ended, Hill and Griggs were free to concentrate on other enterprises. Hill focused on railroad building. Griggs entered into association with Addison Foster.

Addison Gardner Foster

Addison Foster, like Chauncey Griggs, was of old New England stock.[9] He was born in Belchertown, Massachusetts on January 28, 1837, to a family that had been in the area since the early 1600s. His parents soon headed for the frontier and most of his childhood memories were of the white pine woods of Wisconsin and the farmlands of Ohio and Illinois, of Indian scares and ox-cart rides, of pitching hay and gathering corn, and of falling asleep to the thump of the family loom as his mother wove cloth for the children's clothes.

He was about eight when his father took a preemption claim along the Sheboygan River in Wisconsin, not far from Lake Michigan. Young Addison was initiated to hand logging as the family cleared the land. He never willingly touched saw or ax the rest of his life.

Hay pitching he did not mind. His brother Everett delighted in recounting a tale about a time when Addison and another brother, Sam, were putting up a stack on a patch of prairie by the Sheboygan. A party of Indians landed nearby, secured their canoes, and went hunting. The temptation was irresistible. Sam and Add took two of the canoes out for a paddle. The current was too strong. Add couldn't get back upstream so he left the borrowed craft on the bank more than a mile from where he found it.

When the Indians discovered that one of their canoes was gone they surmised that the Foster boys were responsible. Old Iron Thunder, their chief, appeared at the farmhouse, stern and angry. Add confessed and led the Indians to the canoe, which was somewhat the worse for wear. Mrs. Foster promised that the boy would be properly punished.

One evening a few days later Mrs. Foster and Addison were going to the garden patch for vegetables. Old Iron Thunder stepped from the woods and, pointing at Addison, shouted, "Papoose! Canoe!" "Mother was much frightened," Everett recalled, "but she retained her wits. Without hesitation she answered, 'Ah,' which meant Yes in Indian. 'No good papoose. Me whip him hard.' At the same time she slapped her hands together. This seemed to pacify Old Iron Thunder."

The Fosters had been on their claim a little more than a year when word came from Ohio that Mr. Foster's brother, who had been looking after a farm for his widowed mother and sister, had died suddenly. Mr. Foster and his wife discussed going to Ohio. Word spread that he was thinking of giving up his claim. A stranger appeared at the farmhouse carrying a sack from which he poured onto the kitchen table more than a thousand dollars in silver five-franc pieces. It looked like a mountain to the Fosters, who sold the claim.

They traveled back east in an ox-drawn covered wagon over roads that were first deep mud, later frozen ruts. As they passed through Milwaukee, one of the thills (the shafts between which the ox was fastened) came loose. Add was told to hold it in place until they could reach a blacksmith shop. He lay face down on the wagon and reached underneath. The rest of the family marveled at the sights as they drove through Milwaukee. It was, "Oh, look! Lookee here! Lookee there! Just look at that!" until Addison uttered a protest he was never allowed to forget, even when he was a United States senator: "Oh, help. I can't see a dumb thing."

After three months' travel by ox cart, the Fosters reached Ohio. Life was more settled. For Addison it was a life of farm chores and classes in bare schoolrooms, and after he left school nothing but more chores. When gold was reported in Colorado, Addison and a younger brother insisted on trying their luck. Their father gave them a team and a covered wagon. It was Pikes Peak or bust, and they busted. At Doniphan, Missouri, they ran out of money and hired out as farm hands. By the time they had a grubstake, earlier stampeders were returning with tales of hardship on the trail and slim pickings in the gold fields.

"They tried to discourage us from going on," Foster told an interviewer long afterward.

"But you persisted?"

"No, we gave it up and turned back."

"But that spoils my story. It shows a lack of determination."

"But it's the truth."

The Foster boys used their trail money to rent some farmland, which they worked until both came down with malaria. Add's brother gave up and went back to Illinois. Add caught on as a teacher in a small school near Harrisonville, Kansas, at twenty-five dollars a month. After two terms he decided teaching was a bigger bore than farming.

Add's father had accepted some land in Kansas, sight unseen, in payment of a debt. Add decided to look it over. He walked two hundred miles to inspect the farm, only to learn that his father had swapped it for land in Iowa and a note of credit. So he walked to Iowa, found the note valueless, hiked back to Kansas and forced the men to redeed the original land to his father. He then returned to the family farm in Illinois and gave his father the deed. But on the way home he had passed through Wabasha, Minnesota, south of St. Paul. He liked it so much he went back to live there.

Everybody in Wabasha took to Add Foster. His figure was Falstaffian. His laugh was memorable. "It began," someone said, "like a poorly primed pump and ended like thunder in the hills." He told jokes well but was even better at listening to jokes. He remembered names. He enjoyed being helpful. He was decently but not overwhelmingly educated. He was a Mason. Foster had all the basic requirement for political success—everything except burning desire. He accepted appointment as county surveyor, a position for which he qualified by trustworthiness rather than training, but after marrying Martha Wetherbee in 1863, he abandoned public service for a business career.

Foster sold cord wood to Mississippi riverboats, dealt in grain and real estate, opened a freight forwarding business in Lake City and Red Wing, communities halfway between Wabasha and the Twin Cities. Business took him often to St. Paul. There he met Chauncey Griggs, who had served with Everett Foster in the Third Minnesota.

The friendship formed between Addison and Chauncey amounted to another brotherhood. They remained best friends all their lives. Foster moved to St. Paul in 1876. A year later he formed a partnership with Griggs for the warehousing and wholesaling of provisions in the Duluth area, which was expanding under the impact of railroad development. Griggs's old partner James Hill had taken over the St. Paul and Pacific and was completing the spur to Winnipeg. The Northern Pacific, which had gone bankrupt at the beginning of the Panic of 1873, had reorganized and was pushing rails westward. The North Wisconsin Railway was laying steel. "Steel rails now gridiron Minnesota," crowed St. Paul's *Daily Globe,* "and Dakota clasps hands with the commercial emporium. . . . It brings an empire to our very doors."[10]

Railroads required coal and wood for their locomotives, lumber and iron for their tracks. They opened huge areas for agriculture. Farmers needed equipment, farming towns needed everything. Griggs and Foster grasped chances as they arose.

In the winter of 1878 the North Wisconsin, laying tracks southward from Duluth, reached a body of water that the Chippewa Indians called Che-wa-cum-ma-towan-gok, the Lake of the Beavers. Construction workers called it Beaver Dam and named their camp-town after it, but the president of the North Wisconsin decided to call the place Cumberland, honoring his birthplace in Maryland. Whatever its name, the countryside was lovely, a white pine forest interspersed with open fields easy to cultivate. Winter temperatures might be minimal but many of the railroad workers were recent arrivals from Scandinavia and North Germany,

Addison G. Foster (Minnesota Historical Society)

accustomed to cold. They wanted homes, and land could be claimed for settling on it. A farming community grew up around the railroad town.

Griggs and Foster decided the Cumberland area would boom. They bought some land on a peninsula that was almost an island in Beaver Dam Lake and opened a general store, which they supplied from their Duluth warehouse. J. F. Miller, who had mercantile experience, was brought in as manager. The store, officially called the Miller Mercantile Company, soon came to be called "the Company Store" and was so known until it was destroyed by fire in the 1960s.

Foster spent several months each year in Cumberland, watching the flow of goods and looking for new business opportunities. Many of the farmers calling at the Company Store offered to pay for their purchases with hand-hewn lumber—cordwood, ties, and pilings—which theoretically came from timber they cleared from their own land, though the settlers were most casual about the ownership of the trees they felled, feeling that they were doing a man a favor to let a little light into his forest. The Company Store resold the lumber to the railroad.

In 1880 two experienced lumbermen, J. C. Maxwell and T. P. Stone, bought an estimated one hundred million board feet of standing timber from the North Wisconsin Railway land grant and built a sawmill at Cumberland. Foster and Griggs had never been in the lumber business, but their dealings in cordwood gave them knowledge of its potential. The Cumberland operation looked promising.

White pine was called "the monarch of the north," the wood most in demand for lumber. Not only was Beaver Dam Lake surrounded by timber but it was connected by a network of streams to surrounding lakes, the most important of them being Kidney, Sand, and Waterman, which made it possible to tow logs to the mill most of the year and skid them on sleds across the ice in winter. Foster and Griggs bought into the operation. Stone and Maxwell was reorganized in 1881 as the Cumberland Lumber Company, operating as such until 1888 when Griggs and Foster bought out their associates, reorganized as the Beaver Dam Lumber Company with a capitalization of $200,000, and bought another fifty million board feet of standing timber from the land grant.

They were truly lumbermen now, but they continued to operate the Company Store and to deal in local real estate. Two areas of Cumberland are still known as the Foster Addition and the Griggs Addition.

While accumulating white pine stumpage in Wisconsin, the friends were gathering agricultural land in the Dakotas. Griggs's steamboat operations on the Red River made him aware of the farming potential of the area. The Dakota land, level as an ocean becalmed, stretched westward toward infinity. The prairie sod was difficult to break, but settlers who turned it over were elated with the richness of the soil. Cattle had pastures horizon-wide; prairie grasses offered more hay than a man could cut. Markets were distant, but in the late 1870s the Northern Pacific was driving steel

Foster and Griggs's first mill: the Beaver Dam Lumber Company plant at Cumberland, ca. 1880
(*from* The Cumberland Advocate, *1898; courtesy of Howard Jacobson*)

straight through the heartland of North Dakota and feeder lines were sprouting from the trunk.

Under terms of its land grant from Congress, the Northern Pacific was originally to receive alternate sections, twenty miles square, on each side of the right-of-way. Before construction began some of the land had been claimed by individuals, so generous lawmakers gave the railroad an additional five square miles on each side of the track. The railroad possessed eight million acres in a fifty-mile swath through Dakota Territory.

The Panic of 1873 had driven NP stock, issued at one hundred dollars a share, as low as ten dollars. During the post-bankruptcy reorganization of the line it was possible to buy blocks at even lower prices. When the railroad's land company began selling in the Dakotas, it asked two to six dollars an acre, but would accept stock at par value in payment. For those who had picked up NP securities when times were bad and faith low, as had Foster and Griggs, it was possible to acquire potential wheatland for as little as twenty cents an acre. They began buying land in 1880, first as individuals, then as Foster and Griggs Company. In all, they bought nearly 100,000 acres.

Most of their land lay in what came to be called Foster County and Griggs County. The names are a strange coincidence. Foster County was named for James T. Foster, federal immigration commissioner to Dakota Territory, a man Addison never met. Griggs County was first called Ole Bull County, then renamed for Captain Alexander Griggs, Chauncey's associate in Red River steamboating, who was honored as the founder of Grand Forks and as a member of the North Dakota constitutional convention.

Their land speculation and even the Beaver Dam lumber operation were incidental to the Foster and Griggs operations in the St. Paul area. Through the 1880s St. Paul was their base.

In 1883 Griggs formed a grocery firm under the name of Glidden, Griggs and Company. Glidden soon retired and the firm became Yanz, Griggs and Howes. Howes was bought out, Yanz died, and Griggs, Cooper and Company emerged. It became the largest wholesale house west of Chicago. Griggs then organized the Sanitary Food Manufacturing Company, which produced foodstuffs under its own label for sale through Griggs, Cooper and Company.[11]

Foster had a token investment in the grocery business but he was a full partner in the Lehigh Coal and Iron Company, a fuel venture that was started in 1884 and disposed of in 1887 when the peak of local rail construction had passed. Griggs and Foster invested in prospecting ventures in Dakota and Montana without spectacular success. Colonel Griggs visited Spokane and Portland in 1884 to check on mineral possibilities in the Far West, but he did not invest. Both men speculated extensively in St. Paul and Minneapolis real estate of all kinds—city lots, residences, commercial property, industrial buildings, good investments in an area

The Griggs, Cooper and Company warehouse in St. Paul during construction (courtesy of Corydon Wagner, Jr.)

where transportation lines clustered. Foster owned the warehouse occupied by Griggs, Cooper and Company. Each served on the boards of national banks in the Twin Cities.

The friends were active in politics, Griggs as a Democrat, Foster as a Republican. Griggs, who had served as state representative and state senator while living in Chaska, had two more terms as representative and one as state senator from St. Paul. He put in seven terms on the city council. As a representative he introduced the Enabling Act which permitted St. Paul to construct a water system, as a councilman he led the fight for appropriations, and as chairman of the appointive St. Paul Water Commission he supervised construction of the water system which still serves the city.

Foster put in a term on the city council, then devoted himself to building the Republican political organization in Minnesota. In 1882 he was manager for Isaac Stephenson's successful campaign for the United States Senate and S. P. Snyder's run for Congress in the Fourth Minnesota district. Foster was a key figure in the election of L. F. Hubbard as state governor, and subsequently persuaded Dar S. Hall to make a successful bid for Congress. He was also active in support of Senator McMillan of Minnesota in all of his campaigns.

The partners served on various ad hoc committees for the city and county and in numerous trade associations; they were both active in the Congregational Church, and were members of various establishment clubs. In 1882 they built adjoining houses on Summit Avenue, the hilltop boulevard,

Griggs, Cooper and Company warehouse, St. Paul (from a company letterhead, 1901; courtesy of Corydon Wagner, Jr.)

which, with its commanding view of a magnificent bend in the Mississippi, was becoming one of the most fashionable streets in the Midwest, a showplace for the socially elect to display the solidity of their domiciles, the sheen of their horses, and the elegance of their carriages.

C. J. Johnston, a twenty-three-year-old graduate of the new school of architecture at Massachusetts Institute of Technology, was commissioned to design the two houses and a joint carriage house, his first large assignment. He started with the carriage house, designing a handsome building of two stories with a mansard roof of heavy slate. It cost $12,000 and served as workshop for the artisans while the residences were under construction.

The Griggs residence at 476 Summit was designed in the dominant Romanesque style. It was of rough-cut stone with many gables and a slate roof. Its twenty-four rooms had, as befitted the home of a dealer in wood and coal, twenty-seven fireplaces, and it cost $35,000. The Foster house next door was of wood and brick, somewhat smaller—only twenty rooms—and more comfortable. Except for round-arched windows, which gave a Romanesque touch, it reflected the Queen Anne style just coming into vogue in the Midwest. Its cost was about $23,000. The lawns and gardens of the two residences intermingled and the Griggs and Foster children looked on both houses as home.

Thus, in 1888 the partners seemed firmly settled in St. Paul. They were recognized community leaders, wealthy and honored, happily married, the fathers of considerable families. Griggs was fifty-five years old, Foster, forty-eight. It was at this settled stage of life that they embarked on a new venture. They moved to a frontier community on the West Coast to start new lives and build a new industrial empire.

The forests of western Washington had interested the lumbermen of the upper Mississippi for some time. Frederick Weyerhaeuser, largest of the midwest operators, briefly considered Northern Pacific land-grant timber and a millsite at Tacoma in 1885, but decided the time was not ripe. The following year he and Peter Musser took an option to buy 80,000 acres but again backed off. With the start of work on the Cascade Division, which would provide a direct connection between Puget Sound and eastern Washington (instead of the roundabout route through Portland), interest quickened in the forests of the Pacific slope.[12]

In January of 1888, on the tenth anniversary of their partnership, Foster and Griggs accepted an invitation from T. F. Oakes, vice-president and executive officer of the Northern Pacific, to travel west in his private parlor car to look at the trees at the far end of the line.

The party was met in Tacoma by Paul Schulze, a handsome man of forty. A protege of former NP president Henry Villard, Schulze had survived Villard's loss of power. Urbane, imaginative, persuasive, and corrupt, Schulze as land agent for the railroad had a hand in an extraordinary range of Northwest projects, but Griggs was not favorably impressed by him, perhaps because of Schulze's extravagances, which included a taste for the fanciest in parlor cars, wines,

The Griggs residence (foreground) *and the Foster residence were among the showplaces on St. Paul's Summit Avenue, "the most fashionable thoroughfare in the west." This photograph was taken about 1905 but the houses had not been altered since they were built (Minnesota Historical Society)*

horses, and actresses. The forests Schulze had to display, however, Griggs and Foster admired.[13] They made several trips by horseback into the high foothills of southeastern Pierce County, a region dominated by Douglas fir. The enormous trees, many of them more than 200 feet tall and some nearly 300 feet, straight-trunked, branchless for 150 feet, deeply impressed the men from the white pine country.

David Douglas, the Scottish botanist for whom the fir was named, described it as "one of the most striking and truly graceful objects in Nature" and predicted that its wood "may be found very useful for a variety of domestic purposes: the young slender ones exceedingly well adapted for making ladders and scaffold poles, not being liable to cast; the larger timber for more important purposes."[14] Griggs and Foster thought in terms of houses for a growing nation.

One day Oakes invited them to dinner at the Tacoma Hotel to discuss business. He added that he wished them to meet two other businessmen who had come from Wisconsin and Michigan on a mission similar to their own, Henry Hewitt, Jr., and Charles Hebard Jones.

Henry Hewitt, Jr.

It would be interesting to know what first impression Griggs and Hewitt made on each other.[15] Both were successful businessmen and they were destined to become life-long friends and partners, but they could hardly have been more dissimilar in temperament and style. Griggs at fifty-five was courtly and portly, soft-spoken, conservatively and expensively dressed, the beau ideal of the successful Victorian businessman.

Hewitt, six years Griggs's junior, was described by a friendly reporter as owning "two million dollars worth of property in his own right but to judge from his dress, manner and appearance, the natural supposition would be that he possessed little else than himself."[16] His suit was rumpled, his shoes down at the heel, his gray-peppered hair unruly, his manner assertive, his grammar uncertain, and his statements—especially with reference to himself—tinged with hyperbole and impish bravado.

Hewitt was born in Yorkshire to a family of Quaker farmers whose ancestors had known influence and prosperity. By his own account his father "was brought up on a farm, to the strictest principles of industry and virtue, but with not more than three months' school instruction during his life." In 1840, the elder Hewitt, when twenty-eight years of age and the father of three children, the youngest being the one-month-old Henry, Jr., sailed for America "to better his condition." In the frontier town of Racine, Wisconsin, he found work swinging a pick. Within two years he was prospering as a contractor on harbor improvements and was able to send for his family.

Mrs. Hewitt (Mary Proctor) was helplessly seasick for almost the entire voyage. Sailors looked after the children.

Young Henry so enjoyed the trip that he sometimes claimed to have been born at sea. Certainly he loved to travel.

His father, too, kept on the move. Having established a reputation in Racine as a competent and reliable builder, Hewitt, Sr., was lured to Chicago as a subcontractor on the Illinois Canal, then nearing completion. The man for whom Hewitt served as agent disappeared with the company funds. After meeting the claims of his workers the elder Hewitt was flat broke. The family went back to the land, farming a claim near Milwaukee, but prospered no more from agricultural pursuits in Wisconsin than they had in England.

When Wisconsin became a state in 1848, the Hewitts moved to the Fox River country. They were drawn by a proposal to connect the Great Lakes with the Mississippi by means of a canal linking the north-flowing Fox with the Mississippi by way of the southward-flowing Wisconsin River.[17] The first white men in the area, Louis Joliet and Father Marquette, had portaged between the Fox and Wisconsin, and for two centuries the route served the fur traders of four nations. Now it was planned to open the way for steamers by digging a canal between the rivers and building locks at the rapids. Congress authorized a grant of half of the land for three miles on each side of the Fox to be used in financing what was called the Fox and Wisconsin Improvement. The project was ill-conceived, under-financed, and awash with opportunism and corruption. There were so many delays that railroads were serving the area before much steamer traffic was generated. Even though the Improvement failed to bring all the benefits anticipated, it did lead to the transfer of much public land to private ownership, which spurred growth in Wisconsin. Waterpower harnessed by dams allowed industrial developments and new towns appeared along the Fox.

Hewitt, Sr., was involved in much of the construction around Green Bay, Oshkosh, Appleton, Kaukauna, Menasha, and Neenah. His first work was at Kaukauna, downstream from Lake Winnebago, where he had contracts to build a canal and a dam. He moonlighted as a lumberman, cutting and hauling pilings needed on the project. The family migrated up and down the Fox as work progressed, and young Henry's schooling was, at best, intermittent. In 1853, when he was thirteen, his father enrolled him at Lawrence Institute in Appleton.

Lawrence eventually became a university, but at the time it had yet to graduate a college class. The institute consisted of a single wooden building of four stories, 20-by-70 feet, which had been knocked together in a community house-raising at a total cost of seven thousand dollars. The official history of the university describes the hall as "a marvel of inconvenience, which held the whole school—faculty, students, classrooms and living quarters—higgeldy piggeldy. The Methodist Church occupied much of the first floor, and it is rumored bed bugs occupied the third. It is known that the fourth was surrendered to bats and ventilation."

Young Henry proved resistant to the offerings in this hall

of academe. He paid a three dollar tuition charge on March 17, 1853, but was not billed for room and board. Probably he lived at home in Kaukauna, six miles away. The junior preparatory form in which he was enrolled offered classes in English, Greek, Latin, and mathematics. His record shows no demerits, a considerable feat for a boy attending a school that forbade its pupils to enter saloons *or* grocery stores, banned "clamorous noise, athletic exercises, and smoking tobacco in the seminary buildings," and "the use of profane or obscene language, intoxicating drinks, playing games of chance or indulging in indecorous conduct at the seminary or anywhere else."

No grades are listed for young Henry either. He may have avoided both temptation and formal instruction by not attending class. He was in later years to claim that he won the contract for building the dam at Appleton. In fact, his father held the contract, but it is not improbable that he was helping his father at the dam site when he was supposed to be in class. At any rate he did not return to Lawrence for the fall semester, though two of his sisters enrolled. Instead he served as a timekeeper for his father. For the next twenty years the Henry Hewitts, father and son, were associated in many projects. It is not always possible to tell which of them was boss.[18]

The day books of Morgan L. Martin, a former congressman who had sponsored the legislation for the Fox and Wisconsin land grant and for a time had charge of the project, give glimpses of the elder Hewitt's work on the Fox and Wisconsin Improvement. He appears first in charge of a work gang on the Kaukauna canal, making minor purchases for his men at the company store. ("June 13, Tobacco & 4 pipes, 1.16¢. Gallon molasses, .50¢. . . . June 21, 1 novelty lock, $2.00. 2 bolts, 2 screws .29¢. . . . June 27, Straw hat, .37¢. July 11, 1 cowbell, $1.00. 7¼ lb. butter, $1.00.") An entry four years later lists payment of $4,990.04 for canal work. Young Henry's name begins to appear in 1855.[19]

The Hewitts moved to Menasha in 1854 when the senior Henry won a contract to enlarge the existing canal, build crib work for a dam, and put up a wooden bridge.[20] The Fox flows out of Lake Winnebago in two channels, which unite at Little Lake Buttes des Morts. Between the channels lies Doty Island, a mile-and-a-half long, a mile wide. The rival towns of Neenah and Menasha stand on opposite sides of the Fox, each claiming a part of Doty. The Hewitts settled on the Menasha part.

The island was timbered with a beautiful stand of hickory and oak, but it was no haven of refinement. A visiting divinity student complained that he had "not seen a Bible since I came into the State, except on the Pulpit cushions. . . . The smokers include everybodie, Men, Boys and I do not know but Women. So much for Wisconsin." A young lady newly arrived from New York wrote home that at night there were only two lights: "One was the sawmill and the other Neenah." A doctor warned his fiancee that "the island is large, covered with trees, with streets turnpiked

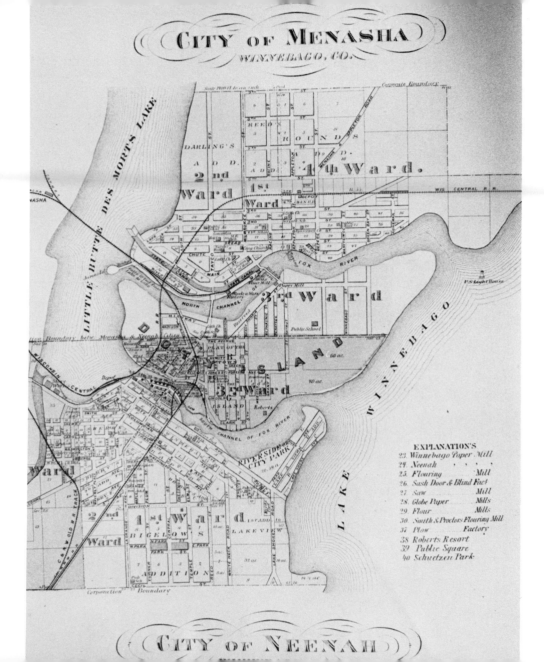

The Fox and Wisconsin Improvement, designed to make a river connection between the Great Lakes and the Mississippi, caused the town of Menasha to boom (Menasha Public Library)

over a great part of it. The streets seem to me much like a new spider's web set to catch houses but none have as yet fallen in except here and there a poor little forlorn midget."[21]

One of these forlorn midgets housed the Hewitts. It grew as Menasha grew, fertilized by the wealth created by public works, and it became a personal citadel on a street named Hewitt. The man for whom the street was named put the proceeds from his contracting ventures into a variety of local enterprises. During the late 1850s Henry, Sr., was owner or partner in a general store, a woolen mill, a flouring mill, a sawmill, and a pail factory.

The elder Hewitt's business instinct was forward-looking—he sought to anticipate demand in a growing community and to command a good credit rating—but his operating methods were notably old-fashioned. John Strange, a native of Menasha who later gained recognition as an inventor and politician, grew up near the Hewitt place on Doty Island and as a young man worked in old Henry's flouring mill. His unpublished memoirs recall with bemusement the way the short, ruddy-faced Yorkshireman went about doing business.

There was no formal exchange in Menasha for the sale of wheat. Farmers from the surrounding counties simply carted their grain to town, usually in late fall or early winter when frozen ground compensated for the lack of surfaced roads. They bargained with the millers in the street. One day Hewitt's partner, Alexander Syme, was indisposed, and Hewitt assigned young Strange to go out and do the purchasing.

"I knew nothing about the quality or kinds of wheat and told Mr. H. so," Strange recalled.

He told me to go on and learn, and instructed me how to tell damp wheat by pushing my hand down into the sack, and also smelling of it. He told me not to buy anything but hard Fife wheat and to pay from $1.02 to $1.05 depending upon how clean it was.

To start out without any knowledge and such a meager advice seemed perilous to me, as my purchases were likely to run as high as $2000 to $3000 during a day. I kept watching during the first day for someone coming from the mill to call me back, but no one disturbed me. I saw long lines of teams driving up to the mill's hopper with the wheat I had bought, and when the day's work was over on the street I went down to the mill with a good deal of timidity for fear I had made mistakes too great to tolerate a continuance of my services as a wheat buyer.

The old gentleman Hewitt told me with smiles on his face that I had made but one mistake during the day and that a small one. He said, "I think you do better than Syme, and we'll have you do the buying on the street in the future." I did this important work for the firm and had a trying time of it.[22]

Strange thought that differences of opinion between Hewitt and his partner, as well as "a too rigid sense of economy," prevented the mill from having such standard equipment as a safe or money box, cash books, or ledgers. "As our cash sales frequently ran to $500 or more after banking hours, I had to carry that money in my possession and take it home with me overnight."

Henry Hewitt, Sr. and Jr., were involved in an extraordinarily varied array of enterprises (Menasha Public Library)

Behind the times though he might have been in safeguarding cash, locked in the mind-set of an old country farmer, Hewitt Senior prospered. He was in a good place at the right time. The Fox country was growing.

In 1856 the promoters of the Fox and Wisconsin Improvement publicized the progress of their project by buying a small paddlewheel steamer, the *Aquila,* at Pittsburgh. They took her down the Ohio to the Mississippi, up the Mississippi to the Wisconsin, through the new canal to the upper Fox, down to Lake Winnebago and then down the lower Fox to Green Bay on Lake Michigan. "Thus," said a local editor, "was consummated on June 16 the marriage of the waters of the Mississippi and Lake Michigan." The Hewitts, father and son, were in the assemblage that joined the Menasha Brass Band in serenading the paddlewheeler as it thrashed through the canal system they had helped create.[23]

Menasha and Neenah prospered from the publicity about the maiden voyage, though the Improvement was far from ready for heavy river traffic. In 1857, old Henry opened a general store in Menasha, "across the street from Bishops Hall" according to his advertisements. It was "the best stocked and most completest" in town. Goods included wallpaper, ladies bootees, fancy silk vests, mechanics' tools, Irish linens of all prices, and "Teas from 31 cts to $1 per lb."

It was in connection with the store that the Hewitts first appeared in the confidential reports of the Mercantile Agency, a credit-rating organization of the day. "[Henry Hewitt] is an Englishman," says the agency man's entry for October 10, 1857, "and commenced rather poor some 3 years ago as a contractor for River Improvement and has been more or less in that Bus. ever since and has made money out of it. Supposed to be worth several thousand. Has a good assortment of stock on hand in store but does not attend to the Bus. himself, leaves it mostly to his Clerks. He also has a contract on the Maintewac [Manitowoc] R.Ry and doing gd. bus. with a gang of hands, and I believe a pretty Shrewd Manager generally. . . . ."

A year later, another report described Hewitt as being "of good character, said to have considerable capital and to own val. r.e. in this and neighboring counties, with little or no encumbrance. Somewhat hard up at present but considered responsible for all his engagements. Think him worth 15 thousand."

Being "somewhat hard up" probably meant that Hewitt was land rich but low in ready cash. This reflected the influence of young Henry, who, while still in his teens, had become a land speculator. Helping his father supply lumber for the Kaukauna project, he had developed a talent for estimating the yield and the difficulty of harvest on forest lands. It was young Henry's good fortune to become a timber cruiser without portfolio when white pine, the monarch of the north woods, was abundant. Soft and durable, easy to nail, paint, or glue, slow to warp, shrink, splinter, or decay, lustrous when planed, white pine made the most popular lumber in the Midwest. It could be had almost for the asking. So much was logged without permission from public lands that the government counted itself lucky when anyone paid a stumpage fee. Many recognized the utility of white pine but young Henry was among the first to realize that the forest would not last forever. He began buying white pine stumpage, blocking it into tracts economical to harvest, and holding it against the day when diminishing supply drove up the price.

The federal government, too, was land rich and cash poor. Holding title to much of Wisconsin and almost everything west of the Mississippi, government agents were eager to transfer land into private hands and get it on tax rolls. A speculator like young Henry could acquire land in many ways. He could squat on a tract, then buy it at minimum price under the Pre-emption Act, or purchase the land from someone who had already gone through that formality. He could buy at considerable discount the military scrip the government issued to soldiers in lieu of pay, then turn the scrip in at face value on the purchase of land. Often it was possible to acquire at bargain rates forest land which obliging appraisers classified as swamp. All states had land that they could sell to raise money for schools. In Wisconsin 680,000 acres had been set aside for sale to finance the Fox and Wisconsin Improvement. In this latter grant, young Henry found great opportunity.[24]

A private corporation had taken over construction work on the Improvement but was having trouble raising cash.

When Henry Hewitt, Jr., was eighteen, he offered to accept payment for some construction work in pine lands rather than cash—if the land was discounted 80 percent. On such terms he won a contract to build one of the dams on the upper Fox. His father helped with the project but the proceeds went to the son, who found himself in actual possession of real estate, the first property he owned in his own name. He liked the feeling. Thereafter, whenever possible he bought pine lands.

The Civil War broke out while young Henry was still working on the dam near Portage. Though professing a longing for military service, be permitted himself to be persuaded that he would be of greater service to the nation as a contractor. He hired a substitute to serve his hitch, a standard procedure during that war, and was amused when he learned later that his father, without telling him, had already sent somebody in his place. Thus he was doubly represented against the rebels.[25]

The war meant new opportunity for men already possessed of plant and capital. Hewitt, Sr., sold his general store but retained ownership or interest in mills processing wool, wood, and wheat, all of which were needed by the military. A labor shortage developed as men went into the army, but women soon replaced them in the mills. The ladies worked a patriotic ten-hour day, sixty-hour week, at minimum wages, turning out fabric for uniforms and blankets, wooden containers to hold military supplies, and flour for food. Wholesale prices rose with wartime inflation. The

Henry Hewitt, Jr., at the time he came west in 1888. He much preferred old clothes (from W. F. Prosser, A History of the Puget Sound Country)

price of wheat doubled, that of wool more than doubled, and that of white pine buckets almost tripled. During the war, old Henry's worth in the estimation of the Mercantile Agency rose from $15,000 in 1859 to $50,000 in 1866, while his habits, character, credit, and business ability, previously described as "good" became "very good."

As for young Henry, after finishing the Portage dam, he concentrated on buying land, picking up mining claims as well as stumpage. When the Confederacy began to crumble and it seemed peace might break out, the elder Hewitt, fearing a business collapse, advised his son to dispose of property acquired for speculation and to squirrel up cash against hard times. Young Henry demurred. He explained later that his years "had not been the long ones of nerve-weakening struggle that had been his father's lot." Instead of selling out, he bought up old Henry's pine lands, rode out the post-war recession, and emerged by his own varying estimates, from $30,000 to $200,000 to the good.

His gains the younger Hewitt continued to invest in more land. The Homestead Act, permitting citizens to acquire title to 160 acres by living on the claim for five years, the Morrill Act, making large grants to the States for the endowment of agricultural and mechanical colleges, and the Pacific Railroad Act, bestowing tremendous land grants on the builders of transcontinental lines, all passed by Congress in 1862, had opened a new empire to land speculators.

Young Henry continued to do his own timber cruising until 1866. On a prospecting trip north of Green Bay he tripped the wire on a spring-gun apparently set for bear.

The slugs caught him in the thigh. He crawled from the woods. Doctors saved his leg, but for the rest of his life he limped. Years passed before he could move without great pain over forest trails. So he took up, among other things, banking.

During the Civil War the elder Hewitt had become a silent partner in the Bank of Neenah, a private bank, whose ownership was a puzzle to the investigator for the Mercantile Agency. When banking laws were revised in 1865, Hewitt surfaced as president of the National Bank of Neenah, which had a federal charter. Although it was capitalized for only $50,000, the absolute minimum under the law, Hewitt's bank received an A rating in Bradstreet's Commercial Report for 1868—the only such rank bestowed on any business in the area. Two years later the Hewitts opened the Bank of Menasha, with the senior Henry as president and son Henry as cashier.[26]

Young Henry later claimed that during his fifteen years as cashier he was mistaken about a loan only once and on that occasion lost the bank only $300. The *Menasha Press,* in a retrospective story about the bank, credited old Henry with being at the bank almost every working hour of every day and making all the decisions. John Strange, the inventor, who dealt with the Menasha bank in those years, described Percy D. Norton as "teller in name but in fact the whole thing at the bank." If indeed Percy Norton ran things, young Henry could take some of the credit. He had hired Norton at the age of sixteen, straight out of Oshkosh High School, starting him on a business career in which he

and Hewitt were associated for as long as Norton lived. Percy married Henry's wife's sister, establishing a relationship they described as "brothers-in-law, once-removed."[27]

At the end of its first year, the Bank of Menasha had loans and discounts amounting to $24,500, almost $14,000 in deposits, a note circulation of $27,000, and total assets and liabilities of $93,000. It never achieved the strength of the Hewitt bank in Neenah, however, and in 1879, to avoid federal taxes, the Hewitts surrendered their national charter. The Menasha institution operated for the next twelve years as a private bank under the name of Hewitt, Son and Company. In 1891 it was again reorganized and incorporated as a state bank, the Bank of Menasha.

Cashiering at the Bank of Menasha was by no means enough to fill young Henry's time. The Mercantile Agency now estimated his worth at $150,000, and that of his father at $200,000. Its agent noted that the younger Hewitt "could undoubtedly draw on his father for what he needs." His credit was "first rate."

In April of 1876, Henry and Reuben M. Scott took the lead in building the first paper mill on the northern channel of the Fox. With his younger brother William, Henry bought the Menasha Wooden Ware Company, and with his father purchased the Menasha Chair Factory, of which he became president. In 1878 he joined two friends and founded the Manufacturer's Bank of Appleton, serving as its vice-president.[28]

Although his leg still gave him some pain, he stumped about the forests of Minnesota, Michigan, Kansas, Arkansas, and Missouri, asking naive questions, wearing clothes nondescript to the point of disguise, while keeping an experienced eye on the prospects for lumber and ore. By 1880 he was reported to have acquired 40,000 acres of yellow pine in Arkansas, and to be holding 1,200 acres in Kansas, 4,000 in Missouri, 8,000 in the iron-bearing region of Michigan, and from 3,000 to 4,000 acres around Duluth in Minnesota.

Henry relished such nicknames as "the plain-clothes pine king," and cultivated his reputation as a curmudgeon. When called on to testify in a suit involving the Fox and Wisconsin Improvement and asked to state his profession, he told the court: "Most everything. My father was a contractor on this canal from the time of its commencement, and I took charge for him and under him. I have been a lumberman; I have been a pineland shark; I have been a tax title thief; I have been a jobber; I have been cashier of a bank. I have got money enough now to get up with most of them. I helped to build a dam at Kaukauna. We drew all the timber, father and I, for it. I have built a good many logging dams on Wolf river. I built the dam at Tobernor's Bend on the upper Fox under contract."[29]

His tough guy posture left him vulnerable to criticism when, during the sharp business contraction of 1883, his Menasha Chair Factory went into receivership. The closure was temporary, but young Hewitt came under bitter attack from the *Neenah Daily Times*. "Henry Hewitt, Jr., stand up," it commanded in an editorial. "If you are not the de-

generate son of a noble sire, you will step to the front and see that these poor men and women are paid every cent the rotten concern you headed owes them. Your noble father never robbed a man of a cent. Will you step forward and maintain the family name and honor?"[30]

Two days later the *Times* apologized by amplification. A story acknowledged that Hewitt had resigned his office in the company months before its collapse, that he was not responsible for the shut-down, and that though not liable legally or morally he was working with the receivers to make sure all workmen were secured in their claims against the company. The workers received their back pay and the plant reopened. Still the attack rankled and the suspicions it aroused smouldered. Henry Hewitt, Jr.'s speculative eye began to rove beyond the Midwest.

In 1886 he sold off $380,000 worth of Wisconsin and Michigan pinelands, pocketed a letter of credit for the whole amount, and set off for Arizona, New Mexico, and Old Mexico, accompanied by his friend and brother-in-law, Charles Hebard Jones.

Charles Hebard Jones

C. H. Jones, a lank, laconic New Englander, was a head taller than Hewitt and five years younger.[31] Born on April 13, 1845, in central Vermont to parents of old English stock (the Jones family in England had lived within fifty miles of the Hewitt, Foster, and Griggs families), C. H.

grew up around saw mills, the water-driven mills of the White River in Vermont, and, after his millwright father moved west in 1851, around the steam-powered mills on the Fox. The Joneses settled in Menasha in 1853, where the father constructed and operated a hub-and-spoke factory. There never was much question that Charles was to sawdust born, destined to cut a lot of lumber in his lifetime, but his folks insisted on some school first. He attended public school in Menasha and went briefly to the preparatory school at Lawrence, then taught grade school for a year.

In May of 1864, at the age of nineteen he volunteered for Civil War duty, though his health was uncertain. He quickly made first corporal in Company D of the Forty-first Wisconsin Infantry, but, when his three-month hitch was up, he dropped out of the army and went back to school, to Ripon University this time.[32] In 1866 he was back in Menasha, helping in the hub-and-spoke works.

Young Henry Hewitt was in Menasha at this time, nursing his wounded leg, and the two men became acquainted. Hewitt was courting Charles's sister, Rocena. In 1868 Henry asked C. H. to cruise timber for him along the Menominee River, which for a considerable distance north of Green Bay forms the boundary between Michigan and Wisconsin. It was an area of superb white pine.

What purchases Hewitt made as a result of Jones's cruise are not known, but C. H. found a home at the mouth of the river. Two lumber towns faced each other across the Menominee where it pours into Lake Superior. Marinette on

the Wisconsin side had an advantage in that most of its mills stood on the bank of the river and were protected by the bar from storms sweeping in from the lake, but the bar made it difficult for schooners to dock. The mills at Menominee faced the open shore of Green Bay. Their log supply was more expensive since the rafts had to be towed across the bar, but the finished lumber could be loaded into vessels that tied up to finger piers jutting five hundred feet into Green Bay. Jones chose the Michigan side.

In 1870, Hewitt and John L. Buell bought out a mill in Menominee and hired C. H. to run it on contract. The old Strauss Mill, as it was called, was a terrible mill and Jones had a terrible time with it (which, as things turned out, was fortunate for Tacoma and the St. Paul & Tacoma Lumber Company). Simon Strauss was an old-timer on the Menominee. He had been a fur trader, grocer and dry goods merchant when, in 1860, he built the second mill on the bay. It was underpowered, poorly located, and the pier he built to deep water was damaged each spring when the ice broke up and crashed ashore in monstrous piles. Strauss could not make money with the mill even in wartime. He gave up after two seasons. The plant had passed through several ownerships, none profitable, when Jones took over its management.

Jones failed, too. There wasn't enough money to repair the machinery. The land behind the mill, on which lumber was stored while awaiting shipment, was a swamp; the *Menominee Eagle* called it "our natural froggery." Rival mills had more accessible piers. About all that Jones got out of his 1870 experience was a realization that a mill needed good machinery, ample space, and convenient loading facilities. Hewitt sold his interest in Strauss's's folly after a year and Jones turned to logging.

1871 was the worst possible year to try logging on the Menominee. From May through September the summer was hot and dry. The needles of the white pines were tinder, and the leaves of the hardwoods drifted down onto moss so dry it turned to powder under foot. The swamps were sere and cracked. The forest seemed to pant. Day after day a hot, dry wind blew steadily from the southwest, soaking up any moisture that was left.

No fire regulations troubled the populace. Settlers burned stumps in their clearings; crews from the Chicago and Northwestern, pushing a line north from Fort Howard to Green Bay, routinely set fires to clear the right-of-way; logging camps ran through the long daylight hours. Numerous small fires broke out during the last week of September. One threatened the village of Sugar Bush and the town of Peshtigo. Families formed bucket brigades to bring water from the river and wet down the rooftops. The danger abated, though in October the sky stayed yellow, the air acrid, and a film of ash covered everything. On Saturday, October 6, the correspondent for the *Marinette Eagle-Star* reported from Peshtigo that "fires are still lurking in the woods . . . unless we have rain soon, God only knows how soon a conflagration may destroy this town."

The next morning the wind rose, and the small, smoldering fires merged into a front twelve miles wide. It swept toward Sugar Bush. There was no saving the village. Inhabitants fled but the inferno caught them on the road, or in the streams where they took refuge, then rushed on toward Peshtigo, which it struck just at dusk. The only safety lay in the river and the townsfolk stampeded into it, the weak being pushed out into deep water where many drowned. A bridge collapsed, a boat overturned.

In the morning the town was gone but the fire was burning miles away. It wiped out the village of Menekaune, ravaged the outskirts of Marinette, jumped the broad Menominee River and damaged sections of Menominee, and destroyed the village of Birch Creek. At Williamsport thirty-five persons roasted to death in a clearing, two others were boiled in the water tank. When searchers reached the spot where a party of Swedes had collapsed while trying to build a fire break, the only trace they could find of the men were some melted axe heads. Before the rains came and the fire burned itself out, 1,250,000 acres were ravaged and more than 1,100 lives were lost. Three hundred and fifty unidentified bodies were buried in a mass grave at Peshtigo alone.[33]

Tragedy created opportunity. There was much rebuilding to be done. In 1872 Charles Jones again took over the old Strauss mill, this time in partnership with Clinton B. Fay. It made enough money so that Jones joined Fay in forming the Exchange Bank of Menominee, which served more as broker's office than bank.[34]

Charles at last felt ready to marry. He took as his wife

Franke M. Tobey of New York, whom he had met when she visited relatives in Menasha. They started to build a house in Menominee in August. In September the business panic of 1873 struck. By December Jones had lost the mill, lost the bank, lost the house, lost everything except his beloved Franke—and twenty-six dollars.

It took him five years working as manager of a stave factory at Dexterville, Wisconsin, and similar jobs to get together enough money to try again. In 1878, with $2,500 and some backing from Hewitt, Jones returned to Menominee and took over the old Strauss Mill for the third time. He found a partner in R. Ramsay of Appleton, who put up money for machinery.

This time the mill prospered. New machinery was installed, the finger pier was lengthened and strengthened. Jones used sawdust and slash from the mill to firm up "the natural froggery" and transform it into a suitable storage area for lumber. To the relief of the townfolk who had for years shuddered under a rain of ash and glowing cinders, the conical wasteburner was rebuilt and rescreened. Ramsay and Jones built a company store, then a commodious, brick business block on Main Street.[35]

In 1884 Jones joined six local businessmen in forming an electric light company, capitalized at $14,000. This was before the days of incandescent lighting. Their equipment consisted of two arc-light generators of about twenty-five horsepower each. The company did well, but in 1891 the Menominee Electric Light, Railway and Power Company was organized with capital stock of $110,000 (of which

Charles Hebard Jones and Franke Tobey Jones, about 1885 (Tacoma Public Library)

"The Tacoma" on the bluff overlooking Commencement Bay and the mud island known as "The Boot" was the most elegant hotel in the Pacific Northwest (from C. Clark, Tacoma, the Western Terminus of the Northern Pacific Railroad)

Jones held a considerable share). The new company bought out the property of the old, enlarged the plant, and built the first electric street railway in northern Michigan, with six miles of track and five cars.[36]

Jones was also a director of the Lumbermen's National Bank and of the Menominee River Boom Company, and a partner in a shoe company. He built a big, comfortable wooden house of three stories and many dormers. He joined the Masons and, on the strength of three months' military service, the Grand Army of the Republic. And he found time to travel, going abroad with Franke and around the western United States with his brother-in-law in search of investment opportunities.[37]

In 1887, the year young Hewitt sold $380,000 worth of his pinelands, the brothers-in-law went to the Southwest to check out mineral prospects. Their trip carried them into Mexico. Impressed by the ores of Sonora, Hewitt built a smelter on the American side of the border at Nogales. He sent Percy Norton, the teller at the Bank of Menasha, to run the plant but simultaneously Congress raised the tariff on imported ores. Hewitt shut down the smelter in 1889 and twenty years later had not found a buyer for the property.[38]

Meanwhile, Hewitt and Jones had visited California in the fall of 1887. The redwoods impressed them but Hewitt noticed that fir was being used more in San Francisco than the local lumber, a circumstance which he decided "did not indicate a particular advantage in redwood for one desiring to engage in the west coast lumber trade." They headed north to look at the region of the Douglas fir, checking out the pine forests of eastern Oregon on the way.

On the Olympic Peninsula north of Grays Harbor, Hewitt and Jones studied the giant cedar and spruce of the rain forest but decided the land presented too many difficulties for economical logging. The fir stands behind Port Gardner on Puget Sound tempted Hewitt (he later returned to them), but the visitors found exactly what they were looking for in the magnificent stands of fir on the western slope of the Cascades in southeastern Pierce County.

Visiting the western headquarters of the Northern Pacific in Tacoma, they met Vice-president Oakes. He suggested that they have dinner and talk business with him and two other lumbermen visiting from the Midwest. So it was that the matched pairs of mill owners, Griggs and Foster from St. Paul, Hewitt and Jones from Menasha and Menominee, got together for the first time in the Stone Room of the Tacoma Hotel.

The Big Deal

OVER BRANDY and cigars, in a dining room looking across the empty acres of the tideflats and Commencement Bay, Thomas Fletcher Oakes talked to his lumbermen guests about freight rates and lumber prices. As executive officer to the president of the Northern Pacific, he was an expert on the subject. He conceded that it posed a great problem for the railroad.

The completion of the mainline from the Great Lakes and the Mississippi to Portland and Tacoma in 1883 had touched off a population boom in the Pacific Northwest. Passenger volume on trains to Washington and Oregon was high—so high, indeed, that some eastern publications were warning unskilled laborers to stay away; there were not jobs enough to go around. Westward movement of goods was not unsatisfactory, but little freight was going eastward.

Lumber shipments were disappointing. The smaller mills on the coast sold most of their cut to local consumers; the larger mills such as the Hanson–Ackerson Mill in Tacoma, Pope and Talbot's mills at Port Ludlow and Port Gamble, and Captain Billy Ruston's huge plant at Port Blakely, were cargo mills, a designation given those that sent their lumber by ship, most of it going to markets around the Pacific rim.

While the Northern Pacific was laying track across the West, mills had sprouted in eastern Washington, Oregon, and Idaho to supply the line with construction material. Once the track was in place, most of the inland mills kept on cutting for local markets, which they could supply at lower prices than could mills west of the Cascades. To generate some lumber shipments, the NP had lowered the rate for freight between Portland and Walla Walla to 68 cents, and between Portland and Spokane to 51 cents, a hundred pounds. This helped a bit, but Puget Sound mills could not compete in the inter-mountain market, and neither Puget Sound lumber nor that from the Columbia was going to the Midwest.[1]

In 1886 the Northern Pacific had let contracts for construction of the long delayed Cascade Division, a line that would run from Tacoma directly through Stampede Pass to the Columbia, eliminating the dog-leg detour through Portland that lumber had to take on its way east. Already a switchback over the Pass carried some freight. The Stampede Tunnel would be finished by summer. Its completion would permit the lowering of rates to the East, but it would still be difficult for Douglas fir to break into the midwest market unless the cost of manufacturing were reduced.

The Northern Pacific was anxious to see a new type of lumber mill created on Puget Sound, one that catered not to the local market, nor to the water-borne trade, but instead a mill that would cut lumber primarily for shipment east by rail. It would have to be larger than any mill yet constructed, with the latest equipment and experienced management, and it would need to have an assured supply of logs. To bring such a mill into existence and generate eastbound freight, the Northern Pacific would offer timberlands from its land grant at very favorable rates. To do the job right would cost between one and two million dollars. Perhaps rather than competing with each other, Griggs and Foster, Hewitt and Jones should pool their resources and build the greatest sawmill in the world.[2]

Oakes was a persuasive man, lean and handsome, of Welsh stock. He knew railroading, having spent twenty of his forty-four years in the business. But he did not sweep the lumbermen off their feet. They wanted to know what the freight would be for lumber to the Mississippi. He acknowledged that he could not promise a definite rate but pointed out that the Northern Pacific would have a stake in the success of the mill.

The lumbermen studied his suggestions in pairs and as a quartet. They made their own estimates of costs and returns. They again visited the forest. They looked at possible mill sites. Oakes and Isaac Anderson of the Tacoma Land Company, the NP subsidiary that handled land development at the terminus, tried to sell them on an area midway between Old Tacoma and Point Defiance, a spot called by the Indians Cho-cho-cluth—the place of the maples. It was a narrow bench of land with deep water on one side, and on the other, a slanting clay cliff cut by a shallow draw. C. H. Jones didn't like it.

Recalling his problems on the open shore of Green Bay at Menominee, Jones argued that the site was too close to the cliff. There would be inadequate space for storing lumber awaiting shipment, for shuttling freight cars, or for expanding the plant. Though he was assured that Puget Sound never froze, he remembered the damage done to the old Strauss mill by the ice pack on the Great Lakes and he worried about north winds sweeping down Colvos and East Passage. Even without ice the winds would rock ships tied to the wharf and create loading problems. Rails would have to be extended along the waterfront to reach Cho-cho-cluth, opening up sites for rival mills. He favored the undeveloped tideflats across from downtown Tacoma.[3]

The area that most appealed to Jones was known to Tacomans as "The Boot." It was a low island off the main tideflats, bordered by two branches of the Puyallup River and Commencement Bay. There was nothing above ground on the boot except a couple of squatters' cabins and the remains of an Indian burial canoe. Nor was there anything beneath

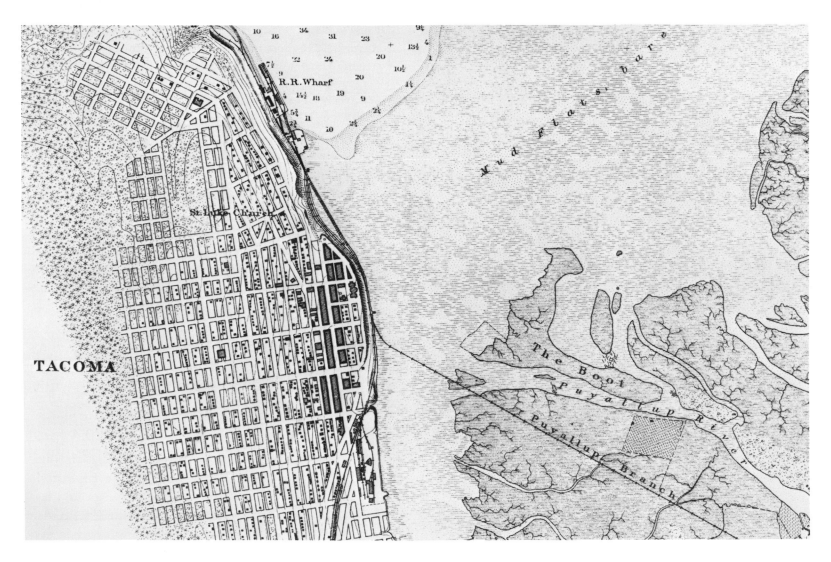

Commencement Bay in 1886, showing "The Boot" and the rest of the tideflat area before filling and bulkheading changed the shape of the Tacoma waterfront (Coast and Geodetic Survey chart; Map Section, UW Library)

the surface except hundreds of feet of silt and dirt brought down from the mountain by the Puyallup, mixed with rotted vegetation. Awash at high tide and when the river flooded, the flats were a great marsh, an expanse of reeds and rushes, ponds and bogs that served as a feeding ground for migratory birds on the Alaska flyway.

Frederick Law Olmsted, the landscape architect who drew up a visionary plan for Tacoma in 1873, only to have it rejected by the land company management, had thought of the tideflats as a potential industrial area; but to most residents, especially those who had hunted ducks and swans in the marsh, the idea of a sawmill out there was amusing.[4]

Why, a mill on the boot would need water wings. Build a hull around it and you could sail it back to dry land or tow the mill to the trees instead of the logs to the mill. And so on. Jones was as firm in his opinion as the boot was squishy, and he was the experienced mill builder in the group. Hewitt backed his partner's judgment all the way. Griggs and Foster were eventually won over by the argument that ample space was a basic requirement. If the deal went through, they all agreed, the mill would be on The Boot.

The lumbermen fanned out on individual missions. Hewitt went to the Coeur d'Alenes in Idaho to look at mining properties, Jones to southern Oregon and northern California to study sawmill operations in the big timber, Foster back to St. Paul to line up other investors, and Griggs, accompanied by Oakes, to New York for discussions with NP President Robert Harris and other rail officials.

While in Manhattan, Griggs conferred with a wealthy cousin, Henry A. Dimmock, who agreed to invest in the project but warned that the Northern Pacific was financially shaky. He urged Griggs to obtain "the best counsel that could be had in New York" to help draw any contract with the railroad, and recommended Francis Lynde Stetson. Griggs secured his services.

The midwesterners regrouped in Tacoma in mid-February for further discussions. The proposal, as it stood, called for the purchase of between 80,000 and 100,000 acres of timberland. The new company would construct not one but two mills, a year apart. They would also build a standard-gauge railroad running southeast from the Cascade Division tracks into the forest. They decided the plan was feasible, took an option on a millsite on the boot and delegated Griggs to return to New York to conclude negotiations with the Northern Pacific.

The first word that something big was afoot leaked into the Tacoma papers on February 17, 1888. The *News* carried a small item on page four under the heading "The Syndicate." It mentioned the presence in town of Griggs, Jones, and Hewitt and referred vaguely to "plans of great importance which on maturity will be given to the public." The two-paragraph story concluded with reference to "a rather inflated item" in the *Astorian,* which quoted Henry Hewitt as saying he was part of a group that had made the Northern Pacific an offer for seven townships (91,280 acres) "and if same is accepted they will build two mills in Washington Territory."

There followed nearly two months of suspenseful silence, broken on Saturday, April 14, by a dispatch from New York in the *Tacoma Ledger*.

TACOMA SOUTHERN

ANOTHER NEW RAILROAD LEADING FROM TACOMA

TO TAP RICH TIMBER & COAL FIELDS

GREAT MILLS TO BE ERECTED ON THE
FLATS OF COMMENCEMENT BAY

New York, April 13. Vice President Oakes of the Northern Pacific says the sale of 80,000 acres of timber land in Washington Territory within 30 miles of Tacoma has been consummated and that papers will be signed tomorrow. The transaction invokes the construction of a line from Tacoma to be known as the Tacoma Southern, to cost two million dollars. The purchasers are O. [sic] W. Griggs, of St. Paul and H. H. Hewitt, of New Richmond, Wisc. and their associates.

In a front page editorial, publisher and editor Rex Radebaugh called the story "the most important item of news that has been printed in Tacoma for several months." He pointed out that the proposed railroad would "tap not only the finest body of timber in the world but also coal fields of unlimited extent and some good agricultural land." He added that "it is almost impossible for those who have not been through the region traversed by the new road to form any idea of the immense quantity of timber to be found there."

The *Tacoma News* the next day agreed that "the value to Tacoma of this new enterprise can hardly be over-estimated."

The contract was not signed the next day. For more than two weeks Tacomans were left to wonder if another mirage was fading away. The most promising rumor came from the *St. Paul Dispatch,* which ran this item: "It is said both Col. Griggs and Addison G. Foster propose to leave St. Paul and take up their residence in Tacoma. If this is so, the loss to our city will be heavy indeed, and it will not be so healthful here, for there is a tonic in Mr. Foster's very smile."

There were no smiles in New York. The proposed deal was in danger of falling apart in a dispute over the mill site. Paul Schulze, the NP land agent, had gone east for the final negotiations. He still favored the Cho-cho-cluth site and persuaded most of the NP directors that a mill on the boot would founder in the mud. They tried to get the lumbermen to accept the twenty acres on which the smelter was eventually built, but Jones would not agree. The NP people then proposed two mills, one at Cho-cho-cluth, the other on the flats. Jones and Hewitt wired Griggs to get the tideflats or break off negotiations. The Northern Pacific at last agreed to let them put their mill where they wanted it. The railroad donated forty acres of the boot for the project and sold an additional 160 acres for $50 an acre, making the total purchase price $32,000. (Eighteen years later St. Paul & Tacoma sold 40 acres of this land to the Union Pacific Railroad for $671,582.) After Attorney Stetson put the contract "in more precise wording to protect the purchasers," the papers were signed by Griggs and Robert Harris on May 2, 1888.

Schulze wired the news to George P. Eaton, chief clerk in his Tacoma office: "Contract was executed today between the company and a syndicate of Minnesota and Wisconsin lumbermen involving the sale of the timber on eighty thousand acres of land in Pierce county, the construction of forty miles of railroad, the erection of two large sawmills at Tacoma, etc. Operations will be started at once. This enterprise will add largely to the population of Washington territory and the enhancement of her material interests. You are authorized to give this publicity. I start for home on Saturday morning."

Eaton rushed to Rex Radebaugh, the publisher of the morning paper, with the news. The next day's *Ledger* gave the story two-and-a-half columns on the front page under the headlines

IMMENSE SAW MILLS

TO BE ERECTED AT THE HEAD OF THE BAY
NEGOTIATIONS WITH NORTHERN PACIFIC BOARD OF DIRECTORS
SUCCESSFUL
AND THE CONTRACT SIGNED

MINIMUM CUT TO BE NINETY MILLION FEET PER ANNUM
SEVENTY-FIVE MILLION FEET TO BE SHIPPED BY RAIL YEARLY
GOOD FOR ADDITIONAL POPULATION TO TACOMA
THIS YEAR OF 4000

The facts were almost as encouraging as the journalism. The contract called for the purchase of 80,000 acres— 50,000 acres already surveyed by the railroad's timber cruisers, 30,000 unsurveyed—at three dollars an acre. The purchasers pledged to erect within a year a mill capable of cutting 30,000,000 board feet of lumber annually, and a second mill with double that capacity within two years, allowing for a total capacity of 90,000,000 board feet a year.

Under the contract, the Northern Pacific promised that its rates on finished lumber shipped eastward to areas north of the latitude of St. Louis "shall be as low as from Portland to the same points." The mill owners pledged that they would ship at least 75,000,000 board feet on the Northern Pacific each year, provided freight rates permitted them a profit of 15 percent on the actual cost of lumber loaded in Tacoma when sold in the interior and Mississippi Valley markets in competition with lumber from Minnesota and Wisconsin. If the freight rates did not permit the Tacoma mill to make such a profit, they did not have to cut any lumber from the land grant but had the option of taking up to 40,000,000 board feet a year for sale anywhere they chose.

This portion of the contract added up to a promise by NP officials to try to open the trans-Mississippi West to lumber from Puget Sound, a breakthrough, which, if achieved, would accelerate industrial growth in western Washington, especially in Tacoma.

The mill owners agreed to build a standard-gauge railroad

extending into the woods for forty miles southeast of Orting. The Northern Pacific, for its part, agreed to deliver logs from the woods at a cost of a dollar a thousand board feet, using NP equipment, but with the lumber company responsible for loading and unloading the freight cars. Payment for delivery would not be in cash. The Northern Pacific would hold its freight bills for delivering the logs until such time as they equalled the cost of constructing the logging railroad. They would then use the bills as payment for acquiring the line as part of their system.[5]

A few days after the signing of the contract in New York, a reporter for the *Oshkosh Times* of Wisconsin caught up with Henry Hewitt ("next to Senator Sawyer, the richest man in Winnebago county") at the Oshkosh railroad station. He said that Hewitt estimated that the land being purchased contained 2,500,000,000 board feet of timber, "chiefly red cedar, fir and spruce and is situated near the base of Mount Tacoma." On that estimate, the company was paying roughly 50 cents a thousand board feet. Hewitt figured it would cost $20,000 a mile to build the railroad, and $400,000 to construct the two mills. The price of the timberland at $3 an acre would come to $240,000, plus interest on delayed payments.[6] (Under terms of its land grant, the NP paid the government 1¼ cents an acre for the 1,866,363 acres it qualified to claim in Washington Territory.) The in-sight costs of the mill project came to $1,440,000.

A Mill Is Born

TACOMANS, accustomed during the construction of the Northern Pacific to long delays between the announcement of a project and its commencement, let alone its completion, watched with surprise and delight as the new people went about their business.

The Griggs–Hewitt party arrived on the afternoon of June 4, 1888. Before the day was done they had filed articles of incorporation with the Pierce County auditor for the St. Paul & Tacoma Lumber Company, capitalized at $1,500,000. They had held an organizational meeting and elected a board of directors. Colonel Griggs was to be president of the company; Foster, vice-president; Hewitt, treasurer; and George Browne, in whose office the organizational meeting was held, secretary.

George Browne served as a link between St. Paul & Tacoma and the Northern Pacific: he was a nephew of NP Vice-president Oakes. A native of Boston, whose forebears had arrived in Salem from Lancashire in 1635, Browne had been educated in New York.[1] At twenty he went into the Union army, was commissioned lieutenant in the Sixth Independent Horse Battery, and served with conspicuous gallantry. He was frequently cited in dispatches, the first time after action at Kelly's Ford on the Rappahannock during the battle of Chancellorsville. ("The guns were served with great difficulty, owing to the way the cannoneers were interfered with in their duties," wrote General Pleasanton. "Carriages, wagons, horses without riders and panic-stricken infantry were rushing through and through the battery, overturning guns and limbers, smashing caissons and trampling horseholders under them. While Lieutenant Browne was bringing his section into position, a caisson without drivers came tearing through, upsetting his right piece and seriously injuring one of his drivers, carrying away both detachments of his horses, and breaking the caisson so badly as to necessitate its being left on the field.") Browne was again mentioned after action at Cedar Run. He rode with Sheridan on the raid to cut Lee's communications with Richmond during the battles of Wilderness and Spottsylvania, and was in action at Yellow Tavern near Richmond when the Confederate cavalry genius, Jeb Stuart, was fatally wounded. For all his gallantry, he was only a senior lieutenant when mustered out.

After his war experiences, the young Yankee saw little risk in Wall Street. A big man with bold, open features, a cascading mustache, and wide cultural horizons, he accumulated friends and market tips with equal ease. In his mid-forties he decided he had made enough money and took his family to France for an extended stay. On returning to New York in the spring of 1887 at the age of forty-eight, Browne was invited by his uncle Thomas Fletcher Oakes, to cross the country to Tacoma to attend a Fourth of July celebration that would mark the completion of the temporary switchback railroad over Stampede Pass.

The party reached Tacoma a day early, and Rex Radebaugh of the *Ledger,* invited Oakes for a drive in the country. Rad, a lean, long-nosed man with a trim vandyke, had more in mind than fresh air and a look at the Mountain. He hoped to interest the NP vice-president in local real estate. They drove south past Wapato lakes where Rad had a homestead (he yarned about the time he sat on his porch and watched a bear kill a calf that had become bogged in the mud between the lakes) and went as far as the pleasant rise known as Fern Hill. When Oakes admired the terrain, Rad revealed that he had an option to buy at $75 to $100 an acre some two hundred acres from homesteads owned by two Civil War veterans. Would Oakes care to join him in a development?

Oakes agreed on condition he could bring in his nephew, George Browne, "who came with me from the east and has a room at the Tacoma Hotel and plenty of ready money. I have the highest esteem for him and want him to locate in Tacoma because I have great faith in the town." That evening, back at the hotel, Browne accepted the proposition without even looking at the land, and Tacoma acquired a highly useful citizen.

When Oakes returned east, Browne stayed in Tacoma to work with Radebaugh on real estate promotion. They called their development the Oakes Addition, using the name "as additional proof that the railroad company continued to stand firmly back of Tacoma, thus correspondingly discouraging Portland and Seattle boomers."

To make the addition accessible Radebaugh and Browne incorporated a streetcar line, which they called the Tacoma and Fern Hill. It did not go as far as Fern Hill, but Browne, a nature lover, liked the name. His brother, J. Vincent Browne, an engineer, came to Tacoma to supervise construction of the line.

The partners took a suite of three rooms on the second floor of the Campbell and Powell Building, 936 Pacific Avenue, two blocks from the *Ledger* office. They brought in two rolltop desks and, according to Rad, "went together into that specialty of the real estate business which is limited to the division of acres into town lots." When Hewitt, Griggs, Jones, and Foster made their first visit to Tacoma in January of 1888, Browne met them, became interested in their plans, and promised to invest if a company were formed.[2]

So it was that when the parlor car "Glacier" arrived in Tacoma on June 4, 1888, Browne was on hand to meet the

George Browne, ca. 1890 (from H. Hunt, Tacoma, Its History and Its Builders)

St. Paul & Tacoma party. The offices of the Tacoma and Fern Hill Street Railway Company became the temporary headquarters of the lumber company. Griggs, Foster, Hewitt, and Jones moved into back rooms looking down on Commerce Street (then known as Railroad Avenue), while P. J. Salscheider, the Green Bay millwright, set up his drawing board under a skylight in an interior room. Visitors passing through to see the company officers could pause to watch the plans for the mill take shape.

The Puyallup River had not yet been rechannelled. The stream divided into a wavering Y near the present sewage treatment plant. What was called the West Passage followed approximately the route of today's main channel, but the East Passage, which was then the main stem of the river, curved across the tideflats and reached salt water at Thirteenth Street on what is now called City Waterway. Isolated between the two channels and the bay was the glob of muck known as "The Boot."[3]

In trying to fit a sawmill onto the boot, Salscheider's first problem was getting his men and material across the river. His solution was a trestle. It was put up in a hurry. Three days after the company was incorporated its surveyors were taking sightings on the tideflats.[4] On the seventh day, they did not rest, they maneuvered a pile driver into place on the East Passage. In two weeks the piling was in place and tracks crossed the Puyallup to the wasteland.[5]

Hewitt and Jones had located a small sawmill at Hot Springs near today's Lester on the NP track to Stampede

Pass. It had supplied lumber for railroad construction but now was idle. They bought the mill, had it taken apart and freighted to Tacoma. On June 20 one crew of workers laid a plank walk from the rails to the millsite "so as to keep anyone from being lost in the mud,"[6] while a second crew carried the temporary mill into place. No one disappeared. The workers put up a building 35-by-70 feet, which stood at the southeast corner of the boot, less than a yard above high tide.[7]

Hewitt went out by rowboat to inspect the site on July 11. He told reporters he was delighted at the progress. Piling was being driven for the foundation of the main building of the permanent mill. Carpenters were completing the frame for the portable. The machinery—boilers, circular saws powered by eighty-horsepower engines, carriage planers, pulleys, shafters and belters—were on site, ready for assembly. Other crews were building a three-story boarding house, 49-by-32 feet, with a two-story addition at one side. Hewitt predicted that within a month the portable mill would be in production, and it was.[8]

By mid-August the portable was turning out planks for flooring, boards for siding, and timbers for the frame of Mill A, which had the proportions of a football field, 300-by-66 feet. Masonry was supported by pilings 16 to 18 feet long, which were driven in until their tops were flush with low tide. The areas that supported machinery rested on solid masses of piling. The 166 longest and thickest piles were clustered under the gang saws. The base for the gang,

22-by-26 feet at the bottom, 14-by-20 at the apex, was flush with the low-tide line. Tacomans, watching the pile drivers at work in the bog, were still not sure that the mill wouldn't slowly subside into the silt when the machinery was installed.

In addition to the main plant, Salscheider's plans included a 44-by-45-foot shingle mill, two engine houses of stone and brick, each about the size of the shingle mill, and a drying kiln 115-by-90 feet. C. H. Jones designed a network of canals and holding ponds in which logs could be stored, then floated to the saws.

While the plant was taking shape on the boot, the organization of the company was completed in the makeshift boardroom. Land was purchased for lumberyards in Yakima and Ellensburg, and for a headquarters building in Tacoma at South Twenty-third and Adams (now Holgate). A strip of tidelands 500 feet wide connecting the mill with the open end of the tidelands was leased from the Tacoma Land Company. A call of 10 percent upon the capital stock was issued to pay for the improvements, subscribers receiving a new share for each one hundred dollars paid in.[9] Percy D. Norton was elected assistant treasurer and was entrusted with the company books. Norton became one of St. Paul & Tacoma's most popular figures, both with his co-workers and with the general citizenry, who elected him repeatedly to the city council.[10]

The mill owners planned to live in Tacoma and be active in community life. Within a week of their arrival, Griggs

The Griggs residence at North Fourth and Tacoma Avenue about the time of completion
(*from* Spike's Illustrated Description of the City of Tacoma, *1891*)

and Hewitt bought residential property on Buckley Hill, the high ground northwest of Old Woman's Gulch, where the Tacoma stadium was later built. The hill commanded a spectacular view across the sound to the Cascades and the Olympics. James Buckley, division superintendent for the Northern Pacific, had planned to use the entire block bounded by Tacoma Avenue and E Street between Fourth and Fifth streets, but when offered $2,000 a lot, agreed to sell Griggs four lots fronting Tacoma Avenue and Hewitt four fronting E Street.[11]

Colonel Griggs had brought with him plans drawn by C. J. Johnson who had designed his St. Paul mansion. They called for a wooden building of three stories, plus a basement containing kitchen, pantry, storage, cellar, laundry, boiler room, and fuel room. The basic structure was to be fifty feet square, with verandahs looking out to the east and south. Andrew H. Smith, a Tacoma contractor, was hired to supervise construction according to Johnson's plans and work began at once. The residence was almost ready for occupancy when a reporter for the *Tacoma Globe* visited it in April 1889.

The newsman spared few familiar extravagances in describing the spacious entry hall from Tacoma Avenue with its elaborate fireplace, the finely furnished reception room divided from the main hall by intricate wood screens and portiers, the spacious dining room overlooking the bay, the large and well outfitted butler's pantry, the library with its redwood bookcases and mantels, the "colonel's very own

sleeping apartments, which are amply provided with closet, bath, and toilet rooms," and the lavatory off the vestibule with all modern conveniences.

The second floor with a hall and five large chambers brought forth leaner prose, but on the third floor the visitor marveled at "the spacious hall with windows facing three sides of the building, from which one can view the grandeur and beauty not alone of the peaceful waters but the snow-capped summit of Mount Tacoma." The billiard room opened onto a covered porch over the front entrance. The mansion was steam-heated, plate-glassed and "furnished with wires for all necessary call bells, burglary alarms and electric lighting." An elevator ran from the basement to the top floor. Why, the place would cost $20,000 when complete.[12]

The St. Paul & Tacoma headquarters at Twenty-third and Adams (Holgate) was completed the same month. It was a two-story building with a tower surmounted by a tall flagstaff. The ground floor contained a general store, which stocked dry goods, groceries, boots, shoes, and crockery from which the camps in the woods and the boarding house beside the mill were supplied. The store was open to the public as well. The headquarters held several suites for the officers, and a reporter for the *News* noted that "desks are surrounded with elegant hardwood rails, which make the place look less like a lumber company office than a bank." Not so surprising when one remembers that Griggs, Foster, and Hewitt all were bankers or bank directors. The com-

The Hewitt residence at North Fourth and E streets was under construction when Rudyard Kipling visited Tacoma and wrote of "the cattlemented, battlemented bosh of the wooden Gothic school." This handsome building has been replaced by a parking lot (from Spike's Illustrated Description of the City of Tacoma, *1891)*

pany safe had a time-lock and was said to be burglar proof and fire proof.[13]

The first leg of the company railroad, the Tacoma Southern, was completed the same week as the headquarters. The first standard-gauge logging railroad in the country, it ran south from Orting into the timber for ten miles. Two logging camps had been established in the woods, cutting had begun, and the flatcars were rolling about thirty carloads of logs a day into the storage ponds on the boot.

Boilers at the mill were fired on April 15. A week later, April 22, 1889, Mill A went into production.

"WHIRR-R, WHIZZ-ZIP," read the headline over the *Ledger*'s account of "glistening steel saws that seem hungry to slice up great fat logs; keen-edged planes that ground and floated over the planks and boards fresh from the humming saws; gigantic band saws that hissed as they glided through high piles of flooring and joists like the thin blade of knife through soft cheese; ponderous pulleys and shafts—all moving in harmony and driven by two magnificent 275-horsepower engines. . . ."

All the founding fathers of the company were on hand in business suits and rubber boots that overcast Monday morning to accept congratulations and answer questions from dignitaries and reporters. George Browne served as master-of-ceremonies, explaining that Jones and Salscheider, who had designed the mill to take advantage of its watery location, were understandably too busy at the moment to talk about their achievement.

"The cheapest way to handle logs is by water," Browne went on. "The secret of profit in the lumber business lies in reducing the amount of hauling to a minimum. The branch railroad tracks to the Boot are arranged so that the logs can be dumped into the ponds just as they come from the woods. The ponds are filled by the tide and locks hold the water in at low tide, so the logs can be floated up to the mill whenever needed."

As Brown spoke, the first Douglas fir log, deep barked and nearly a yard in diameter, was brought to the bottom of the loading ramp where it was caught by one of the dogs (hooks) on an endless chain that ran up the center of the trough-shaped slide. The log was pulled up to the saw room, which stood on the upper level of the mill, while the heavy shafting, machinery, and pulleys occupied the lower level.

Steam-driven arms rolled the log into position on a platform facing the band saw, a loop of steel 56 feet in circumference, 10 inches broad, its serrated cutting edge honed to razor sharpness. This was the first time the band saw, literally the cutting edge of sawmill technology, had been used on logs this big. There were doubts that the thin steel could survive contact with the massive, dense, pitch-pocketed fir. The principle of the band saw was not new. It had long been used in England and France for light work. More recently the Hammond brothers of Fort Wayne, Indiana, had built a saw that would cut walnut, and Dolbeer and Carson in Humboldt County, California, had used band

Sketch of the first St. Paul & Tacoma plant (*from* Northwest Magazine, *March 1889*)

saws on the enormous (but soft) redwood logs. Salscheider himself had installed the first band saw used in Wisconsin on white pine. Like C. H. Jones he believed it would work here. He had spent weeks adjusting the main pulleys, 11½ feet in diameter, which, driven by belts 42-inches wide, spun the bands. He was confident but nervous.

The log was pinioned to the platform and the platform slid forward pushing the face of the log toward the whirling steel. The hum of steel cutting air changed to the tortured scream of steel biting into fir, but the blades moved unhesitatingly through the log, a thin stream of sawdust pouring from the kerf. The lumbermen studied the dust intently; one of the advantages of the band saw was that it would waste less of each log than the circular saws, which consumed as much as a half inch of wood per slice.

Salscheider relaxed, almost smiled. Jones remained impassive, intent on the log. In less than three minutes the fir was reduced to slabs that moved along steam-driven rollers toward the lath mill, to boards and planks that went to the planing machines to be smoothed into flooring or other house lumber, or to the gang saws to be chewed into scantling or joists. Defective lumber was shuttled to the shingle mill. The sawdust was carried to burners in the engine room. The planed boards, casing and other finished wood moved on rollers to the yard to be piled for delivery or bunkered up for transport to the drying kiln.

Nine-foot band mill (from West Coast Lumberman, *1904)*

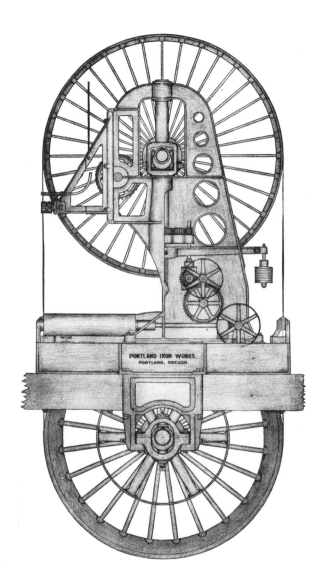

The kiln was a building 115-by-90 feet, nearly air tight, heated by steam from the boilers. Cars with the newest, greenest lumber were placed at the greatest distance from the steam pipes. Each day they would be brought closer to the heat. After six days a load could be at the hottest part of the kiln without warping or cracking.

Everything was working as planned. The mill, at this stage, could produce 190,000 board feet a day. In time, Mill A's capacity would rise to 250,000 board feet. Billy Renton's mill at Port Blakely had cut a record 320,000 feet a day, but when the second St. Paul & Tacoma mill went into production, the complex would be the world's greatest lumber producer.[14]

The partners returned to their new headquarters to drink a toast and plan the future.

Douglas fir was the basic tree in the company's forest land (from West Coast Lumberman)

"Now You're Logging . . ."

THE FIRST section of the railroad that the St. Paul & Tacoma promised in their contract with the Northern Pacific had been completed shortly before Mill A went into production. It marked a new stage in the industrialization of logging in the Pacific Northwest.[1]

In the early days on Puget Sound, logs were taken from stands adjacent to salt water or large tributary rivers, rolled into the water and floated to the mills. Even shoreside logging posed problems of movement. The enormous firs and cedars of western Washington were too heavy to be hauled by the carts and wagons used in New England and the Midwest. They had to be dragged by teams of oxen along skid roads—paths of notched logs set in the dirt at right angles to the log being moved. This was slow work involving several men and up to a dozen oxen for each turn of logs. It was uneconomical to haul logs more than two miles along a skid road.[2]

By the early 1880s lumbermen were running out of forest within skid road range of water. "Logs that are easily shot into the large streams and the once fine bodies of timber along Puget Sound and the Columbia River are even now almost a thing of the past," declared *West Shore* in 1882. The answer, the magazine declared, could be found "in the vast untouched forests of the mountain watersheds [that] can only be reached by logging railroads."[3]

An early line, probably the first logging railroad in the Northwest, was run from the forest around Tenino to salt water at Olympia in 1881. The track was narrow-gauge, the roadbed uncertain, the rails given to sudden dislocation, but the operation was profitable. Logs could be moved farther and faster and cheaper than by oxen. The following year, five more railroads reached from the sound to the stands of rough-barked fir and cedar in the high hills. Steam beat sweat as a means of getting logs to the saw.

In 1883 the Port Blakely Mill Company, long a leading force in northwest logging, began laying track westward from the sound, a well-built line which they hoped to extend to the forests bordering Grays Harbor. Railroads of this caliber cost so much that only companies with a secure hold on the stumpage it intended to cut could afford the investment. The day of the big-time operator with vast holdings of timberland was dawning. By 1888, seventy-five

The replacement of ox-teams by the steam locomotives marked the transition of west coast logging to a capital intensive enterprise (Tacoma News Tribune)

to eighty cars a day clanked down to the Port Blakely mill bearing up to 225,000 board feet of logs.[4]

The St. Paul & Tacoma's line was the next stage of development. It was a broad-gauge line, the first to be built for hauling logs, and built to fit eventually into the regular NP system. H. S. Huson, a division superintendent for the Northern Pacific, engineered the road; gangs that had helped build the Cascade Division laid the track; NP locomotives and flatcars carried the logs. (In 1898 the Tacoma Southern officially was absorbed by the transcontinental.)[5] Its first purpose was to assure a steady flow of logs from camps in the 80,000 acres that St. Paul & Tacoma had purchased.

Hewitt and Jones made the final choice of the site for the first camp—cookhouse, bunkhouses, blacksmith shop, and barns—some five miles up the track from Orting. They also picked the spots for the landings to which logs would be brought for loading onto flatcars. When the landings were established, skid roads were built back into the timber.

Constructing a skid road in the Cascade foothills was no small engineering assignment. The roadbed had to be carefully graded if the logs were not to hang up, causing costly delay. Once the surface was smooth, skids were made from small logs, a foot to a foot-and-a-half in diameter, ten feet long, and notched in the middle. These were placed across the road about seven feet apart. Setting the skids required skill. They had to rest solidly in the soil, for they would bear tremendous weight, and the notches had to be at an even plane so the butt of the log being moved would not nose down against the skid and knock it out of place. Where the skid road curved, the skids had to be set with the outer ends lower than the inner to help guide the turn of the log.[6]

With Camp One built and the landings in place, St. Paul & Tacoma was ready to start logging. Each man in camp had a specialty, and the responsibility of performing a difficult, often dangerous, job fast enough so as not to hold up the crew members back of him. The foremen never let anyone forget that time was money and they were being paid handsomely—from $1.50 to $1.75 for a ten-hour day.[7]

Actual logging started with the faller and his helper. Their task was to drop the trees without shattering them, as close as possible to the skid road. Throwing a 250-foot Douglas fir or cedar accurately through forest tangle in a country that was built with much of the land standing on end called for prodigies of judgment. Fallers had to allow for tree, terrain, and wind. They controlled the direction of fall by undercutting on the side they wanted the tree to lean toward, then sawing from the opposite side. When the bole was nearly cut through they drove steel wedges into the kerf. A good man could drive a peg with his tree. After dropping several giants near a skid road, fallers would move to another road, perhaps a hundred yards away, making room for the sawyers to work on the fallen.

Sawyers usually worked in pairs. The foreman told them the lengths required. Their concern was to prevent the saw

Fallers usually worked in pairs priding themselves on the synchronized rhythm of their strokes and the smoothness of the notch. High stumps, and the notches for the springboards, indicate trees felled before chain saws (from author's collection)

from "running"—going off at an angle. If in cutting a log four or five feet in diameter a saw was off six or eight inches from right angle, the loss in cuttable lumber was considerable.

After the sawyers came the swamper. He cleared windfalls, brush, and branches from the ground between the log and the skid road. If there was time he sometimes went in ahead of the fallers to remove salal, vine maple, devils club, and other torments.

The barkers came next, their numbers depending upon the time of year. In the spring, when the sap was running and bark came off easily, two or three men could do the work, but later in the season it was necessary to have larger teams. Next to the enormous size of the trees, which scaled out ten times as big as midwestern evergreens, the thickness of the bark was the hardest thing for newcomers to accept. They were accustomed to trees with bark so thin it scraped off when a log was dragged. Here, a mature fir had bark eight to ten inches thick. An unpeeled log simply would not glide. The barkers cut through the bark along the top of the log with broad-bladed barking axes, then used barking irons—an instrument like a crowbar, with one end flattened and bent for prying—to work it off down each side.

When the log was properly naked the hook-tender took charge. He decided which side would ride most easily, then snipped the nose of the log at an angle to lessen the chance of its bumping a skid log. A log well snipped, pulled onto the skid without the swamp hook being jerked out of place,

It took about two hours for a bucker to saw through the log from a nine-foot tree with an eleven-foot saw. Lumbermen estimated that old Douglas firs averaged a century of age for each foot in diameter (John D. Cross, photographer; from author's collection)

The donkey engine speeded the movement of the logs from their place of growth to the railroad siding. Machinery was expensive but, unlike oxen, did not have to be fed when not working (Tacoma Public Library)

then riding easily on its proper side, was the proof of a hook-tender's virtue.

The hand-skidder's job was to get the log onto the skid road. He set small logs, five or six inches in diameter, close together along the path the log was to take. Once the log was barked and snipped it was rolled onto its riding side. A block and tackle was brought in from the landing by the horse team. The block was fastened to a tree, the tackle hooked to the log, and the horses slowly pulled the log to the skid road. Sometimes it took three or four resettings, and an amplitude of obscenities, to get the brute out of an awkward position.

With the log in place on the skids, the team was hitched directly to it. Horses were used rather than oxen in the first St. Paul & Tacoma camps. A team consisted of from four to ten horses, depending on terrain and size of logs. The horses were big, well-trained, and well-fed. The forty-four that St. Paul & Tacoma used in the woods averaged 1,525 pounds and burned enough hay to make the switch to donkey engines profitable in 1893.[8]

When the team shouldered into their harnesses and leaned into the load, a greaser went ahead of the log dabbing the notches in the skids with a swab dipped in dogfish oil. (The job was not only the worst paid but the worst smelling in the woods.) Some sang chanteys as they skipped down the road with the peculiar hopping gate that made a greaser recognizable at a distance even upwind. Many greasers were Indians and one of the work songs was in the Chinook Jargon. "Chickamin, luckatchee, klootchman, lum!" sang the skidder as he splashed the rancid oil. It translated into a poor man's dream: "Money, clams, women, whiskey."[9] On the way back into the woods from the landing, the greaser was supposed to sweep off each skid so it would be uncluttered for the next log.

At the landing an NP supervisor directed the loading of logs onto the cars by St. Paul & Tacoma workmen. They used a large jack screw to start the logs rolling down a ramp onto the car and heavy wooden wedges to hold the bottom logs in place. Those on top were stacked without retaining cables. Spills were not infrequent as the logging trains moved down the tracks. Nobody claimed that logging wasn't a dangerous business.

From twenty-five to thirty carloads a day came down from the mountains in the summer of 1889.[10] A year later the rails had been extended another six miles and more camps opened. Fifty carloads a day were being spilled into the waiting ponds on The Boot.

To Market We Must Go

"A man must sell his ware after the rates of the market."
—*John Ray, 1670*

GETTING LOGS from the forest to the mill proved to be easier than getting lumber from the mill to market, especially the midwest market. The company's contract with the Northern Pacific contained the railroad's guarantee that rates from Tacoma to the area between the Rockies and the Mississippi, north of St. Louis, would not be higher than rates from Portland. The catch was that these rates remained too high to make Pacific Northwest lumber broadly competitive with white pine from the Great Lakes and the Mississippi Valley.

In the year between St. Paul & Tacoma's purchase of its timberlands from the land grant and the start of production at Mill A, the price of common lumber in the Puget Sound region fell 9 percent—from $11 a thousand board feet to $10—and the skid did not stop there.[1]

The Northern Pacific, which had its own problems, insisted it could not reduce rates.[2] The fact that the railroad slid into the hands of receivers in 1893 indicates that its claims may have been valid, but the rate structure did put the operations of a mill conceived with the idea of midwest sales in an awkward situation.

The price of the best grade of flooring in Kansas City was $25 a thousand. To ship southern pine flooring to that market cost 22 cents a hundred pounds; to ship Douglas fir flooring from Tacoma, 55 cents. That amounted to a difference of $9.60 a thousand feet, which was an enormous handicap to overcome in introducing a product (Douglas fir) whose superiority in performance could only be demonstrated by widespread use.[3]

Nor were freight rates the only problems that Griggs, Foster, Hewitt, Jones, and Browne had to ponder when they gathered in the second-floor boardroom of the headquarters building. There was a chaos of free competition. With prices falling after a long period of expanding investment and production, with the market tightening at home and abroad, with banks insistent, many lumbermen faced failure. There was price cutting, of course; that was to be expected. But there was also deception as to quality, quantity, and delivery date. The expedients resorted to by men whose alternative was bankruptcy undercut the reputation of stable operators seeking to build an enduring market.

Colonel Griggs and his partners agreed that the short-term solutions to their problems would be advertising, promotion, and the building of a sales organization. In the long run they favored self-regulation of the industry, not only for quality but for price.

St. Paul & Tacoma made the routine short-term moves. A sales team was formed and an ebullient salesman, Emory "Siwash" White, was put in charge of the Midwest. Samples of Douglas fir lumber were sent to every major lumberyard in the country, endorsements were solicited, advertisements run. Sales were made and a market established—but always there was a cowbird around ready to benefit from the efforts of others. The problem was summarized in a letter from Tacoma published in the *Puget Sound Lumberman.*[4] It was signed "Reform," which may well have been the *nom de guerre* of Colonel Griggs, for it reflects his prose style as well as his attitude.

Satisfactory results cannot be attained in the car trade branch of the Puget Sound lumber business nor its possible development reached under any conditions until some radical changes are made in the manner of treating those who are disposed to buy for shipment into the treeless section west of Chicago and east of the Rocky Mountains.

Very few concerns in the whole of the Puget Sound country are in a position to handle, with any degree of promptness, such orders as can be secured for shipment by rail. They are so few that they can be counted on the fingers of one person. . . . It is safe to say that a large majority of the orders for the better grades of lumber—no matter if they are for no more than one carload—are not shipped within thirty days. Add to this four or five days consumed in transmission of mail and probably three weeks for the shipment to reach its destination and it makes a lapse of sixty days from the date of making the order to that when the buyer gets the stock. . . .

A trade of any considerable proportions cannot be developed in this manner with yard dealers located at such a great distance from the point of shipment.

A great deal of good, effective advertisement has been done for the fir, spruce and cedar of the Pacific Coast in the past two years, and the excellency of these woods for the purposes for which each is best adapted has been sounded far and wide. As a consequence numerous letters of inquiry have been received by manufacturers and dealers. The probability is that in all cases these are answered promptly and in a lucid and businesslike manner, and are closed with a strong statement as to the correspondent's excellent facilities. The inquirer is encouraged and sends a trial order. If it is such as trial orders usually are it is for a mixed lot of five or six different items that the buyer naturally supposes is in stock ready for shipment. This is a point on which his informant has taken care not to touch. If the buyer has the faculty of patience largely developed his order will be filled with lumber that is quite sure to meet his expectations; he finds the price lower, is satisfied with all of his experiment except the long delay and concludes to try again, this time impressing upon the party

Special timbers of almost any size and length could be cut from the giant logs. For nearly half a century St. Paul
& Tacoma could supply perfect specimens of 12-by-36 inch, 40-foot stringers (from American Lumberman, *1921)*

from whom he buys the absolute necessity of prompt shipment, etc. The same delay is experienced with this as with the first shipment, and the strong probability is that he gets disgusted and does not care to put any further dependence on getting any kind of stock that can be secured elsewhere than Puget Sound.

Thus the trade of one who might have become a large and continuous buyer is lost. If this loss affected only the party that made the long delays it would be all right and as good as he deserves but the whole of the fraternity suffers from timberland owner to dealer.

That is not the worst of it, for the buyer who has been imposed upon takes no pains to disguise the fact. On the contrary, if he has the average lumberman's combative disposition, he loses no opportunity to tell his experience, and thereby gives the business of Puget Sound shippers a black eye at every turn.

Eastern buyers are partly to blame . . . for in many cases they accept the lowest price offered {and} go to some "broker" whose stock-in-trade exists wholly in an office desk, a chair, a showy letterhead, some postage stamps and unlimited nerve. The sooner both buyers and producers understand they are being imposed upon by a class of so-called "dealers' without responsibility or anything in their business of a tangible nature to commend them to consideration from either buyer or producer, the sooner the buyer will have a better opinion of the coast lumbermen and the manufacturer satisfaction in the car trade.

It is not reasonable to suppose that a manufacturer who has sufficient capital to build and operate a sawmill and appurtenances necessary to handle car trade with satisfactory promptness is going to encourage a broker in his own locality and as far removed from the trade as the manufacturer himself is by paying him a commission, and giving prompt attention to business that were it not for the existence of the desk, chair, stationery and nerve would come to him direct. As a consequence orders placed with local dealers (that are simply dealers) are usually handled by small sawmill operators with little capacity for making such lumber for eastern markets, and less knowledge than capacity. In such cases poorly manufactured and badly graded stock is added to long delay, and again the reputation of the whole lumber business suffers.

The time has arrived when the buyer should post himself as to who in this country are responsible manufacturers and who are little better than parasites, and a time when shippers should adopt the plan of telling the truth about when accepted orders will be filled.

While there is no doubt but that a reduction of rail freights and a consequent reduction of delivered prices would greatly increase the demand for car shipments east, without a disposition on the part of shippers here to stick more closely to the truth concerning their ability to fill orders, the reduction in freight rates and consequent increased demand would probably increase the number of disgusted dealers at the other end of the line to such an extent that the good effect of the lower rate would be lost in bad advertisement. In the lumber business, as in any other, honesty is good policy.

To combat the activities of the parasitical and incompetent, Griggs favored the creation of an industrial association such as those that had set standards and prices on coal back in St. Paul. A first move was the establishment in Tacoma, early in 1891, of a Lumber Dealers Protective Association.

No record survives of what its members agreed to, but their failure to hold rank did not go unreported.

The *Puget Sound Lumberman* carried a story that at an August meeting of the Lumber Dealers Protective Association, the Hart Lumber Company of Tacoma was accused of cutting prices. George E. Hart cheerfully acknowledged that his company had indeed given bargains. Leigh Walker, its manager, added that every mill in town had been fudging on the tariff, except two. Somebody made a motion to the effect that the Association was ineffective and should be dissolved. The motion carried. So ended the industry's first attempt at self regulation on the sound.

Walker did not name the righteous pair who had been out of step with the price-cutting majority, but the *Lumberman* commented editorially that "one of them is the very concern which could, on account of its wealth and influence, drive nearly if not quite every other mill in the city out of business were it so inclined. But it signed an agreement and adhered to its terms most religiously until as Mr. Walker says nearly every other member broke his word. This mill is the St. Paul & Tacoma, and since it has been victimized by its competitors, it is in turn making life a burden for some of them. Already one mill has shut down and two others are running on half time or less. The St. Paul & Tacoma is getting the trade and by the time that company is ready to sign another agreement for the maintenance of prices, some of its competitors will be glad enough to stick to it."[5]

While waging a price war against its opposition, St. Paul & Tacoma kept offering the peace of regulation. An exploratory meeting of influential lumbermen was held in Tacoma in October 1891,[6] and when business leaders from around the state gathered in Seattle on November 17 to discuss Washington's participation in the Chicago World's Fair of 1893, a Lumber Manufacturers' Association of the Pacific Northwest coalesced.

Percy Norton of St. Paul & Tacoma played the leading role. It was he who introduced the motion limiting membership to those who actually manufactured lumber from logs, thus excluding not only non-producing dealers, but shingle makers, whose problems were separate. The membership fee was set at ten dollars. Management of the Association was left in the hands of the directors; Norton was elected president of the Association and a member of the board.

The Lumber Manufacturers' Association proved no more effective in bringing order to the marketplace than the Tacoma dealers' association had been. Nor did it achieve its grand aim of getting Congress to appropriate funds, or guarantee bonds, for the proposed Nicaragua Canal, "this grandest of all works ever contemplated by the civilized world,"[7] which would open the markets of the Atlantic Coast to northwest lumber. The Association was an idea whose time had not yet come.

It remained dog-eat-dog in the marketplace, and St. Paul & Tacoma was too big and strong to be swallowed by its competitors.

On January 1, 1891, Colonel Griggs was proud to announce "the largest sale of lumber ever made in one order—8,250,000 feet to the Northern Pacific for construction work. During the year the company broke heavily into the market for boxcar flooring and siding, even sending several carloads to Baltimore, though the freight amounted to $27 a thousand board feet.[8] They also shipped to Brooklyn three flatcars loaded with timbers 100 feet long to be used in ship construction—the longest cargo ever moved across the continent by rail.[9]

St. Paul & Tacoma established yards in eastern Washington and at the end of the year invaded the Southern California market through an arrangement with the W. H. Perry Lumber Mill Company and the Pioneer Lumber Mill Company. In that deal St. Paul & Tacoma bought one-quarter interest in the two California companies, paying for the shares in lumber, $61,875 for 375 shares in Perry, $25,000 for 250 shares in Pioneer. The move made St. Paul & Tacoma a major source for lumber in the counties of Los Angeles, Orange, San Bernardino, San Diego, Ventura, and Santa Barbara in California, and in the territories of Arizona and New Mexico.

The year ended with Mill A running two ten-hour shifts a day and with no surplus on hand. During the twelve-month, the company shipped more than thirty-five million feet by rail and almost eight million by sea. "The results of the past year's business," Colonel Griggs reported to the stockholders at the annual meeting in January of 1892, "will be, I trust, very satisfactory, especially considering the dull market for lumber during the past season, the demand having been much less for both foreign and home consumption. Many small mills have closed down and most of the larger ones run on less than half time. Our mill has been in operation full time and while the market price for lumber has dropped from $3 to $4 per thousand from the average price of two years previous, our report shows an actual net profit of $74,762."[10]

St. Paul & Tacoma declared its first dividend: 5 percent on paid-up capital stock.[11]

Booming the Boom Town

A FEW MONTHS after Mill A began cutting lumber, a young British journalist visited Tacoma and reported it a city "smitten by a boom of the boomiest." [1] The digging of the Stampede Tunnel and completion of the Cascade Division, the construction of the mill on the tideflats, the launching of four steamships and several sealing schooners, the extension of streetcar lines to the suburbs, the building of a library, a university, and an opera house, the burning of Seattle, the shift of Washington from the status of a territory to that of a state—these and lesser events during 1889 deepened the conviction of folks on Commencement Bay that they were taking part in the creation of the great city of the West.

Population was estimated at 28,000 in 1889. Not one resident in five had been in town five years earlier. The growth was so sudden, the change so swift, that most Tacomans had seen their town do nothing but grow. They expected expansion to go on forever. A song composed on the occasion of the completion of the Stampede Tunnel and sung at civic ceremonies by a bank president and a quartet of bank clerks caught the tone of boosterism and confidence in the year the St. Paul & Tacoma Lumber Company started production:

> Oh Tacoma, the gem of the ocean,
>   The pride of the North and the West,
> The shrine of each tenderfoot's devotion:
>   The grandest, the noblest—the Best.
> Thy harbors make thousands assemble
>   Thy resources proud to review;
> Thy commerce makes Portland tremble
>   While wheat by the tunnel comes through.

CHORUS

> While the white-winged fleets of the nations
>   Lie anchored in waters so fine
> Each hoping to take on thy rations
>   From farm, forest, field, shop and mine.
> When Bennett and Huson had won it*
>   And light shone from shore to shore

* Nelson Bennett was contractor for construction of the Stampede Tunnel. H. S. Huson had done the engineering.

The N.P. was glad she begun it
  And Portland proclaimed it a bore.

Thy resources vast center hither
  For craft with their banners unfurled.
May the fame we have won never wither
  For our market, you know, is the world.
May the sections united ne'er sever
  But rivals continue to be.
Tacoma, Tacoma, forever.
  The world has our glory to see.
And rivals continue to battle
  Tacoma, the tried and the true.
Seattle, Seattle, death rattle.
  Three cheers, for the tunnel is through.[2]

In the expanding metropolis, opportunity seemed everywhere. Men with capital had only to decide on their areas of
greatest interest. For the most part the St. Paul & Tacoma
leaders followed the advice that Addison Foster gave his
sons: "It is best to deal in commodities, something the people must have, like coal, wood and lumber. The cheaper
you can make the commodities, and sell them, the more the
people will use them. I believe in the magnitude of a business rather than in high profits on a small business."[3]
Henry Hewitt phrased it less elegantly but more succinctly:
"See what the people are going to need, see it first, then get
it, and the market will follow."[4] In their individual as well
as their joint investments in the Northwest, the partners

*The opening of Stampede Tunnel made Tacoma in fact as well as in title
the western terminus of the Northern Pacific (from* The Great
Northwest: An Official Guide, 1890–91; *Pacific NW Coll.,
UW Library*)

concentrated on commodities, transportation, and real estate, including timberland.

George Browne was involved with Rex Radebaugh and T. F. Oakes in real estate promotion and a streetcar line before the others came to Tacoma. When "The Glacier" brought the St. Paul & Tacoma party to town, a gang of fifty laborers had just started grading the right-of-way for the Tacoma & Fern Hill Street Railroad which was to run from Twenty-sixth and Pacific to the Oakes Addition.[5]

Delin Hill, up which the tracks ran, was heavily wooded. Slash from the clearing was dumped in the gulch below (today's Wakefield Drive). On the day it was burned a huge column of smoke rose above the city, bringing hopeful inquiries from Seattle as to whether Tacoma was on fire. By mid-August the grade was finished to the top of the hill and early in November the roadbed reached the Addition. A shipment of rails, sent round the Horn from England in the bark *Melpomene* by Balfour, Guthrie, and Company arrived and track was laid. Browne and Radebaugh bought a little twenty-five-horsepower sawmill that had been used at the portal of the Stampede Tunnel. They used it to make lumber out of trees being cleared from residential lots on the Addition, and the lumber went into houses on the site, thirteen of which were sold by Christmas.

"I hear you're getting into competition with our mill," Hewitt complained jokingly to Radebaugh one day. "I don't think it's fair to treat your friends that way when we're investing so much in your town."

To change the subject, Radebaugh asked Hewitt's advice on the best foundation for the little mill. "With one that size," said Hewitt, "all you do is spike her down on a stump." But to show he was kidding he bought several lots in the Oakes Addition.[6]

The Tacoma and Fern Hill was ready to run on April 4, three days before the deadline set by the city council when it granted the street railway franchise. Politicians, reporters, business leaders, and other freeloaders were invited to ride to the end of the line. Hewitt was off buying another forest for himself, Foster was in St. Paul, Jones was busy tuning up Mill A, but Colonel Griggs found time to accompany Browne on the inaugural run to the boondocks. The occasion was memorialized in some of Rex Radebaugh's most purple prose:

Two horse-drawn omnibusses filled with gentlemen who themselves seemed to be filled with pleasant anticipations started from 1005 Pacific Avenue and ambled gaily down south on the unpaved gravel surface of that noble highway. They crossed the Northern Pacific railroad tracks and continued on their southerly course, bee line, not stopping until they arrived at Delin Street and the temporary city terminus of the Tacoma & Fern Hill Street Railroad.

There on the rails, coupled to a new and shining steam dummy,* polished, painted, varnished, appropriately lettered and

* A steamdummy was a locomotive with condensing engines. In consequence it did not need a blast pipe and in theory was quiet and could be used as a streetcar since it was less likely to frighten horses. The theory was not credited by the totality of urban horses.

pretentious in all ways, dreamily discharging toward the empyrean in customary lazy columns its escaping calories as though enjoying the prospect of showing off before a company so distinguished.

At 3:10 P.M. George Browne's brother, Vincent, who was now manager of the line, declared all aboard and ordered the engineer to "let her go." Exactly seven minutes carried the delighted party up the hill, past a decorated waiting room at Oakes Avenue (now Fortieth Street) and on to the terminus at the sound side of the Addition. The clearing offered a fine view of tanks for the water system, the sawmill and boarding house, a blacksmith shop, machine shop, and other amenities. After a buffet that included wine, salmon, and roast beef the party returned to Tacoma, all proclaiming the venture a success. Theodore Hosmer of the Tacoma Land Company was so enthusiastic he went right out and persuaded an old settler to sell him some adjoining land, proclaiming it Hosmer's Addition.[7]

A year later the Tacoma and Fern Hill lived up to its name. Browne and Radebaugh extended a spur from Thirty-eighth Street to Fern Hill, then on to Edison (now South Tacoma). Tracks were also pushed north from the Delin terminus, across the gulch by way of an intimidating bridge, and along Railroad Avenue (Commerce Street) to Ninth. The final phase saw a spur run from Thirty-eighth to Puyallup. The street railroad system returned no profits from fares but did speed up the sale of property in Oakes Addition and helped develop Puyallup, Parkland, and South Tacoma.[8]

Griggs and Hewitt joined Browne in other investments

The St. Paul & Tacoma retail store stood at the northwest corner of Nineteenth and Pacific Avenues. This picture was taken about 1893 (from Cecil Cavanaugh Coll., Washington State Historical Society)

around town. The summer of 1889 was hot and dry (a long drought contributed to the great Seattle fire) so they formed the Tacoma Ice Company, capitalized at $50,000. They put up a two-story building, 75-by-100 feet, at Twenty-fifth and Adams (Hosmer), and installed what the *News* described as "machinery of the most improved style, capable of producing forty tons of ice daily."[9] The plant required 200,000 board feet of St. Paul & Tacoma lumber.

Browne and Henry Hewitt contracted with the Tacoma Land Company to buy a parcel of land on the boot adjacent to Mill A. They intended to build a sash, door, and blind factory, an enterprise they felt would complement the St. Paul & Tacoma operation. The project was still in the drawing board stage when a midwest lumberman, William C. Wheeler, came to Tacoma for his health. He had twenty years experience in the sash and door business and the opportunities in Tacoma made him feel better right away. When Wheeler met Harry Osgood, another transplanted Iowan who had sash and door experience, they decided to unretire together.[10] When Hewitt and Browne heard that two experienced operators were looking for a mill site they abandoned their plans. They sold their contract to buy the land adjoining Mill A to St. Paul & Tacoma for one dollar. The lumber company then entered into a contract with Wheeler, Osgood, and Company to erect a factory on the site and lease it to them for the manufacture of sash, doors, and blinds, and to supply them with lumber. It was the start of an association between St. Paul & Tacoma and Wheeler, Osgood that lasted for more than half a century.[11]

Hewitt and Jones, in their personal investments, looked to forest land and made extensive purchases around Everett, Chehalis, and Grays Harbor.

Griggs served as contractor for construction of the Tacoma, Olympia, and Grays Harbor Railroad Company's line connecting the Tacoma-to-Kalama branch of the Northern Pacific with the south side of Grays Harbor. He received payment in land and organized the Pacific and Chehalis Land Company, which came to possess 20,000 acres in southwest Washington, and of which he was president.

Griggs developed an interest in shipping, at least partly through the influence of Captain John J. Holland, a lank Vermonter with an Uncle Sam goatee, who had built a small shipyard on the flats next to Mill A. There Holland built a steamer, the *Fairhaven*, a pile driver, and a barge for Nelson Bennett's Fairhaven Land Company, which was booming Bellingham.

Holland also lectured anyone who would listen on the need for a dry dock in Tacoma. "A dry dock built here would pay for itself in a short time," he argued.

After one is built the expense of operating it is slight. Being well situated here it would insure a good income to the man who owned it. We have in the harbor here from ten to twenty ocean vessels every day, and the number is increasing very fast. Now that the Stampede Tunnel opens eastern Washington to us and we are going to export so much wheat, there will always be several in port loading that product.

These vessels come from all parts of the world. Before they leave ports in England, Scotland, Sweden or Norway they are docked and overhauled and repaired. Then they go out with cargoes to some distant port, take another cargo to some other country, and are away from their home port from two to five years at a time. This is the case with many of the vessels that come to Tacoma. If there was a dock here all of the vessels would use it, but they cannot afford the time—unless there is an emergency— to clear for San Francisco or Victoria, the only cities on the Pacific Coast having dry docks.

Why do they need to be scraped and cleaned? Oh, you should have seen the steamer *Olympia* when they took her to Victoria last year. One hundred forty-seven tons of barnacles they got off her bottom. These vessels get so loaded with barnacles sometimes they can hardly sail at all. The captains know that time is money. If possible they'd have their vessels drydocked when they came here.[12]

Griggs was impressed by Captain Holland's proposals. He not only invested $10,000 in a Puget Sound Dry Dock Company, he also signed two notes of $25,000 each to the Merchant's Bank underwriting loans to other members of the dry dock syndicate. William H. ("Billy") Fife, a burly, broad-faced pioneer who had become one of Pierce County's richest men, served as president of the company. H. L. Achilles of the Pacific Store and Warehouse Company, William Blackwell, a hotel manager turned banker, and Rudolph W. DeLion of Port Townsend, were the other officers; Griggs was named a trustee.

Captain DeLion, of French descent but German nationality, was one of the most experienced maritime men on the sound. He had worked his way from cabin boy to captain.

After service in the North Atlantic, DeLion went into the shipping business at Valparaiso, Chile, and Callao, Peru, only to suffer heavy losses during political upheavals. In 1876 he arrived on Puget Sound as captain of the bark *Otago,* which he sold to the Port Blakely Mill. Settling at Port Townsend he engaged in various shipping enterprises and campaigned for construction there of a dry dock, agitation which brought him to the attention of the Tacoma syndicate. They put DeLion in charge of design and construction.

The floating dry dock was built at Port Hadlock at the head of Port Townsend Bay. While the 340-foot long contraption was being assembled, other communities set out to win it for themselves. Seattle and Port Townsend offered the syndicate subsidies to anchor the floating dry dock in their harbors, a proposal not without interest to the officers of the syndicate since the cost of building, which had been estimated at $250,000, was approaching $340,000.

Port Townsend interests actually raised a down payment of $100,000 and turned it over to Captain DeLion. Any chance that the dry dock would be kept in the northern sound ended when a sudden storm drove the huge shell ashore. Damage was minimal, but had there been a ship in the dry docks at the time it would have been disastrous. DeLion and his Tacoma backers decided the dock would have to be moored in the calmest possible waters. The possibilities were narrowed to the mouth of Chambers Creek, south of the Narrows; the Hylebos Creek area of Com-

The S.S. Victoria *rides high and dry in the floating dry dock at Quartermaster Harbor. The hulk in the foreground is the old Fleetwood. Asahel Curtis took this photograph on a glass plate in April 1909 (Washington State Historical Society)*

mencement Bay, or a cove on the eastern shore of Quartermaster Harbor, which was protected by the sheltering bulk of Vashon and Maury islands. The Quartermaster site was selected.

DeLion returned the money advanced by Port Townsend interests though they went to court seeking an injunction to prevent him from giving it back. At 7 A.M. on November 24, 1891, to Tacoma's relief and delight, the giant dock left Port Hadlock under tow of the tug *Tyee*. It passed Port Gamble at 1 P.M. Tacomans assembled on the bluff above the waterfront to look up East Passage for its appearance. Darkness fell with no dry dock in sight. Misgivings grew as a strong wind rose during the night.

The next morning, when the dry dock did not appear, Colonel Griggs, Billy Fife, and other backers of the enterprise chartered the little steamer *Mocking Bird* and set out in search of their errant possession. They met the dock off Bainbridge Island. It had ridden out the storm but the tug was unable to pull her against wind and tide. Now with the sound calm and the tide favorable, the dock, big as a football field, advanced dowager-proud down the East Passage, around Neill Point, and on to the bight now known as Dockton on the east shore of Maury Island. There she was secured to pilings.

Work began at once on a surrounding complex of wharfs, one of them 300 feet long and 32 feet wide, another 120 feet long and 90 feet wide. Each fronted on water deep enough for big ships to come alongside. Two enormous boilers were brought over from Tacoma. While they were being fitted to the pumps in the dry dock, a machine shop was built on the wharf. A work force of fifty men commuted daily from Tacoma on a launch chartered by the company.

The first ship served in the Dockton dry dock was the 170-foot passenger steamer *Flyer*, victim of a fire that had swept her superstructure. The dock was filled with water, the crippled ship floated into place, then the pumps drained the water and the dock rose, with it the vessel. Within half an hour the *Flyer* was high and dry, ready for carpenters, blacksmiths, painters, and scrapers. A new industry had been born on Puget Sound, one of great value to other waterfront businesses, but one that, a few years later, would command far more of Griggs' attention than he had anticipated.

The mineral possibilities of the area also interested Griggs. On his first visit to Tacoma he had been invited to put money into a smelter project that another St. Paul capitalist, Dennis Ryan, was promoting. Griggs accompanied Ryan, Isaac Anderson, and Theodore Hosmer of the Tacoma Land Company on a tour of possible sites for the smelter— Ryan eventually chose Cho-cho-cluth after Jones rejected it for the lumber mill—and he listened to Ryan's talk of a million dollar capitalization and the plans for two water jacket smelters, which would work gold ores from the Concunully district of the Okanogan, which was then the scene of a rush, and from Alaska. Griggs correctly guessed that

The floating dry dock, with the Victoria *aboard, seen from the harbor. The Drydock Hotel at Dockton can be seen to the left of the boiler house (Asahel Curtis, photographer; Washington State Historical Society)*

Ryan's vision exceeded his capital. He did not invest. But when a better-financed group was formed in 1890 to buy Ryan's smelter and enlarge it, adding a sampling works and making Commencement Bay a major port of call for ore ships, Colonel Griggs was one of the backers of the Tacoma Smelting and Refining Company.

St. Paul & Tacoma, as a company, also became involved with mining. It bought half interest in the Wilkeson Coal and Coke Company, an under-financed operation that in 1885 had bought out the mines and claims of Henry Villard's Oregon Improvement Company in the Wilkeson–Carbonado area.[13]

The coal fields in the Cascades produced nearly as much controversy as coal. Strikes were frequent, strike-breakers abundant though endangered, and the labor-management relationship at its most peaceful was war on the simmer. Throughout the 1880s the mines around Wilkeson were plagued by violence. A letter to the *Seattle Post-Intelligencer* described the workers as "a crowd of ruffians, composed of Germans, Italians and Irish . . . the very black vomit of different nationalities—unprincipled as the devil, treacherous as Judas, ignorant as babboons and jabbering like chimpanzees . . . some beastly drunk and the majority half-seas over. . . . [They] besieged the house of the manager waiting for a chance to shoot Mr. Kelly in case he should appear, and then cut off the water so that Mrs. Kelly and her houseful of children might be deprived of it."

An answering letter asked the reader to "suppose these men, such as they are, were imported into this country under the contract labor system of free trade in labor, which the protected interests have adopted for the purpose of enabling these industries to compete with the pauper labor of Europe." Buying into Wilkeson meant buying problems along with potential profit.

Griggs and Hewitt were of different minds about the purchase. Hewitt was wholeheartedly in favor; his motto was to see first what people were going to need and then get it, and coal was a necessity. Griggs, who had experience in the coal business, hesitated not because he doubted that coal was a good investment but because he felt St. Paul & Tacoma should consolidate its position in lumber before branching out. He permitted himself to be persuaded—there were no dissenting votes on the board when the purchase was made—and years later wrote Hewitt to acknowledge that Hewitt had been correct. "I am now very glad that you all insisted upon having the purchase made."[14]

With the takeover of Wilkeson Coal and Coke agreed to, the St. Paul & Tacoma board authorized a call of 5 percent on capital to finance the purchase and increase the coal company's capitalization five-fold to $500,000. T. J. Edmonsen, an experienced engineer, was commissioned to visit Mexico, Central America, and South America, accompanied by George Browne, to study the possibility of forming a new corporation for the export of coal to those areas.[15] St. Paul & Tacoma discussed ways to increase the capacity of the coal bunkers on Commencement Bay, and in an interview with

Coal from the Wilkeson Coal and Coke Company mines was loaded onto ships from bunkers below Old Woman's Gulch, which was later filled in to become the Tacoma Stadium (from C. Clark, Tacoma, the Western Terminus of the Northern Pacific Railroad)

the *News* Griggs spoke of the possibility of using Captain Holland's yard to build a fleet of freighters to carry lumber and coal around the Pacific.[16]

The waterfront on the Tacoma side of Commencement Bay was filling rapidly under the impact of the boom. When the Tacoma Yacht Club held a housewarming for its clubhouse on Maury Island on August 23, 1890, Paul Schulze of the Northern Pacific land department took Mr. and Mrs. Browne and Mr. and Mrs. Hewitt on a tour of the Bay in his naphtha launch *Lillian*.[17] On the Point Defiance side of the smelter they passed the Pacific sawmill which was cutting 150,000 board feet a day. The low, oblong stack of the smelter was pouring dark smoke. Between the smelter and Old Town were two shingle mills, a small shipyard, and two brickyards. Two more shingle mills and a brickyard were under construction in Old Tacoma. The old Hanson–Ackerson mill, now called the Tacoma Mill, lay toward the city, its wharf host to eight sailing vessels.

Down the bay from the Tacoma Mill was a warehouse built by W. S. Ladd and Company, 2,600 feet long, 160 feet wide. It could hold 600,090 bushels of wheat and was a special delight to Tacoma because it had been built with Portland capital. Inboard from the warehouse was a flouring mill capable of grinding 600 barrels a day. From it earlier in the year had gone the first shipment of flour from Puget Sound to China, 32,000 hundred-pound sacks, valued at $72,000, bound for Canton in the *Earl of Derby*.

The Tacoma Warehouse and Elevator, with a capacity of 1,250,000 bushels, and the NP elevator, with a capacity of 500,000 bushels—both built in 1889 to handle wheat brought through the Stampede Tunnel—lay under Cliff Avenue (Stadium Way). They could load six ships at a time. Alongside them were the coal bunkers at the foot of Pacific Avenue and the mill of the Gig Harbor Lumber Company. Beyond that the bay shoaled up.[18]

Across the waterway, out on the boot, St. Paul & Tacoma's Mill A and its outbuildings dominated the tideflats with only Captain Holland's shipyard and the Pacific Naphtha Launch Company dock for company.[19]

Putting the Puyallup in Its Place

THE PUYALLUP River, flowing from ice fields on the shoulder of Mount Rainier, slants steeply westward, gathering snowmelt and rainfall from some 650 square miles as it approaches Puget Sound. For most of its 50-mile course the Puyallup is a mountain stream, swift, tumultuous in freshet, ripping through narrow canyons, frothing over stretches of rounded rock, chewing at bank and bottom in its downward rush. After receiving the waters of the Stuck, however, it meanders across an alluvial delta left by the drainage of a glacial lake that formed during the last ice age. As it nears Puget Sound the river is falling only three feet a mile and the silt and sand it has picked up begin to settle. When the river meets salt water the last of this debris sinks to the bottom and is held in toward the head of the bay by the push of the tides.[1]

At the time St. Paul & Tacoma's Mill A was built, the marshland at the mouth of the river consisted of a 20- to 30-foot layer of silt, sand, fine gravel, and vegetable muck resting on an equally thick layer of clay below which was a deep deposit of gravel on top of the marine sediments laid down in the geologic past.[2]

Near the present Interstate–5 bridge the river divided into two channels. The larger branch—called West Passage on most charts but South Passage on the Wilkes' chart of 1841—looped westward and spilled into a broad, shallow inlet, which ran along the foot of the downtown bluff almost to today's Twenty-third Street. The mouth of the Puyallup pointed straight at the downtown Tacoma hill between Thirteenth and Fourteenth streets. The smaller channel, East (or North) Passage, followed approximately the course that the Puyallup today takes from the I–5 bridge to the bay.[3]

In the month that Mill A started cutting, St. Paul & Tacoma began to drive piles for a railroad trestle running a mile along the 500-foot-wide leasehold that connected the plant with deep water at the bay side of the boot. The bay shelved abruptly at the edge of the tideflats and the wharf balanced at the very edge of the drop-off. Too close. When a cargo of bitumen from Ventura County in California was unloaded on it for a Tacoma street paving project, the wharf collapsed and the whole 325 tons went to the bottom of the bay, never to be recovered. The wharf was rebuilt on longer and more closely set pilings.[4]

While rails were being run to deep water by the mill

owners, the Tacoma Land Company began a major project to narrow and deepen the inlet separating the boot from downtown. Its shallowness had long been a handicap to industrial growth. Nicholas Delin had built a water-driven sawmill at the head of the inlet in 1852 only to have the channel silt up so rapidly that ships could not get to his mill. Delin's lumber had to be rafted, tediously, to ships anchored out in the stream.[5]

The Tacoma Land Company chartered one of the largest dredges in the world to deepen the waterway in 1889. The digger, 120 feet long and 32 feet wide, looking "like a bathtub with a smokestack," had been employed in San Diego harbor. The tug *Vigilante* was dispatched to tow her north. For the trip up the coast, the hull of the dredge was covered with decking, the pumping machinery was disassembled and distributed evenly for weight, the 40-foot stack was shortened, and two casks of fish oil and two of vegetable oil were placed on deck to be thrown over if there were need to calm troubled waters. The oil was not used, though on the 16-day trip north the *Vigilante* picked up survivors of the *Alaska,* which had gone down off Cape Blanco. The crewmen arrived in Tacoma with tales of a group in one lifeboat who sustained themselves by drinking blood caught when they amputated the crushed leg of one of their companions.

The dredge was anchored below Eleventh Street and its machinery reassembled. The 50-horsepower steam engine was fired up in July. Townsfolk lined the bluff to watch the topography of the Tacoma waterfront change. The stern of the dredge was held in place by the spud, a vertical beam 20 inches in diameter and 48 feet long, which was lowered to the bottom. The bow, floating free, could be swung from side to side in a 185-foot arc. The cutting head, mounted at the bow, was a large metal barrel 4 feet in diameter ringed with 6 cast iron knives fastened to a hoop that rotated 15 times a minute. The barrel and an attached pipe were lowered to the floor of the waterway.

Material cut loose by the knives was thrown to the center of the cutting head and mixed with water drawn into the suction pipe. The mix was lifted to a chamber in the dredge from which it was transferred to a 2,500-foot long pipe. A centrifugal pump spewed 19,000 gallons of the mixture each minute onto the tideflats. The water drained off, the dirt remained.

Day after day, around the clock, the dredge chewed up 2,500 to 3,000 cubic yards of submerged silt and sand and muck and deposited it on the shore below the downtown cliff, or on the boot, or on the flats at the head of the bay. Sawdust and other debris from Mill A was added to the mix. Inch by inch the marsh was raised and transformed into land. The area at the head of the bay became firm enough to support the NP roundhouse and yards, and the boot was firm enough to hold the piled lumber. But as the City Waterway was deepened by the dredge, sludge from the Puyallup continued to pour out of the river mouth and settle into the new cut.

The Land Company decided to block off the west channel and divert its waters into the smaller east channel. After consultation with St. Paul & Tacoma officials, who raised no objection to having their island made part of the main, a dredge and pile driver were sent to construct a barricade of dirt and brush, faced with slabs from the mill, across the southwest branch at the Y. The old channel, which had made the boot an island, became a swamp, impossible to walk across but no longer a conduit for the river. Dredging the channel between the bluff and the flats was resumed by the Land Company, and St. Paul & Tacoma began to reinforce the south wall of the remaining channel to prevent the concentrated stream from cutting a path toward the mill.[6]

The build-up of the tideflats and the deepening and narrowing of City Waterway cost $300,000. The work went on for more than two years—and was almost lost in a week.

In late October of 1891 heavy snow was falling in the high Cascades, on some days more than an inch an hour. Rain swelled the lower streams. At the end of the month there was a brief cold snap, followed in November by steady rain. Three inches were reported in Tacoma in twenty-four hours; the fall was certainly heavier in the mountains. After two days the rain slackened but on November 8 a warm Chinook wind added snowmelt to the runoff. The westward-flowing rivers rose toward heights matching the tallest tales of the old-timers.

All day on Monday, November 9, the gray-brown flood rose steadily, ominously, its sound deepening as it churned against the embankments. Great trees, their roots still heavy with dirt, swept past the city and out into the bay; several dead cows washed down, and a barn. As the river rose toward the top of the dam blocking off the Puyallup's old channel, Captain William Hull, manager of the Pacific Naphtha Launch Company just upstream from St. Paul & Tacoma, shut down his plant and set some men to removing launches and machinery from the 100-foot long building, which stood only 20 feet from the riverfront. Other workmen were set to filling sand bags.

Around noon, St. Paul & Tacoma stopped cutting lumber. The men and teams of the day shift began hauling slabs, debris, and bales of hay to the embankment. Under mockingly clear skies on a warm November afternoon they ran a losing race against the rising water.

A bend in the river forced the flood against the slab abutment fronting the Launch Company. Under pressure of the flood, water sluiced through the gaps in the slabs, dissolved the fill dirt behind them, formed a whirlpool between bulkhead and dirt-face, undercut the pilings, pulled loose the slabs.

At four in the afternoon 40 feet of bulkhead and fill broke away with a sound between a slurp and a groan, and moved as a clot out into the torrent, where it disintegrated while the stream chewed into the exposed melange in the gap. As the long northern twilight settled on the tideflats, the river attacked the pilings that supported the foundation of the launch works.

Alongside the factory were the houses of George Hansen, an engineer, and Captain Edwin Purdy, a foreman. These were substantial two-story buildings. As the bank crum-

This photograph by Thomas Rutter was taken in 1891 when ground was being cleared for City Hall. The Northern Pacific headquarters building dominates the cliff; below lies City Waterway, in which dredging has just started. Across the boot stands the St. Paul and Tacoma mill on the old channel of the Puyallup. The mountain has been painted in (Courtesy of Doug McConnell)

COPYRIGHT ©94
BY A.H.WAITE.

The same scene taken by A. H. Waite a year later. The tide is high and the marshland across from the Union Depot is afloat. By 1894 the land had been filled in and the boot was "a piece of the Main." The buildings in the foreground are the fire station and the Tacoma Hotel (Historical Photo. Coll., UW Library) bled, the houses began to settle. Purdy was able with little difficulty to remove his wife and children and even some furniture, but as Hansen started to evacuate his family a great hole opened before the front door. The engineer barely got his wife and baby out through a rear window before the house settled a full story, then, as the waters continued to carry off the dirt below it, slowly rolled onto its side.

Hewitt, Foster, and Griggs came out by boat to study the situation. They conferred with Isaac Anderson, manager of the Tacoma Land Company, and Allen Mason, whose Commencement Bay Land Company held much of the tide-flat area on the far side of the Puyallup. It was agreed all efforts should be centered on saving the dam blocking the old East (South) Channel. If it broke, the $300,000 city waterway would be clogged in a day. The lights of the mill were turned on to help the millworkers as they hauled 16-foot slabs to the endangered area.

During the night a tangle of trees caught on a shoal above the boot and diverted some of the flow away from the dam, but a small knoll known as Snag Island channeled water at an especially weak area. Clarence Bean, a Tacoma Land Company surveyor, took a skiff to Snag Island, planted fifty pounds of dynamite in the obstruction and touched it off, breaking windows in the St. Paul & Tacoma buildings, allegedly causing a relapse in the condition of a woman suffering from pneumonia, but hardly stirring the stumps on the island. At six in the morning, thirty feet of the launch works, which had hung cantilevered over the flood after the foundation washed way, broke off and plunged into the Puyallup.

That was the climax. The dam held, cold weather in the mountains slowed the runoff, and the river began to fall. The loss at the launch works was estimated at more than forty thousand dollars. Damage to St. Paul & Tacoma facilities came to twelve thousand dollars. (Colonel Griggs had little time to celebrate the escape from destruction. On November 17 he received word from St. Paul saying that fire had destroyed the five-story, 330-foot-long Griggs, Cooper and Company wholesale grocery. Damage totalled more than half a million dollars, wholly insured.) A week later the Puyallup rose again after a new rainstorm. The remains of Captain Hansen's house floated out into the bay, some launches swamped, and temporary repairs that had been made on the launch factory were washed out.[7]

In the wake of the floods, Surveyor Bean won acceptance from all parties concerned on a plan to remake the lower Puyallup. Snag Island was removed by stirring up the sand with blasts of giant powder, then deflecting the current against the island by putting in a temporary wing dam and letting waterpower do the work of clearing.

The channel was widened by 40 feet to a new width of 150 feet. The sharp bend near the launch works was reduced to a shallow curve. In this straightening, St. Paul & Tacoma sacrificed three-fourths of an acre, Mason's Commencement Bay Land Company an acre and a half.

An hydraulic pile-driver was brought upriver to set deep

ALLEN C. MASON, PRES.
JOHN E. BURNS, VICE PRES.

FRANK C. ROSS, SEC'Y.
E. N. OUIMETTE, TREAS.

COMMENCEMENT BAY LAND AND IMPROVEMENT CO.

TACOMA, WASH.

H. S. CROCKER & CO. S. F.

"The Boot" before the fill (from the letterhead of the Commencement Bay Land and Improvement Company, ca. 1892; from author's collection)

pilings for a longer, stronger dam across the old West (South) Channel. The empty bed was filled with sawdust, waste lumber, and dredged silt. Thus the boot was firmly joined to the tideflats behind it. West Passage was no more.[8]

The 1892 chart of the Coast and Geodetic Survey shows for the first time the Puyallup River and Commencement Bay in the basic configuration they have in the 1980s.[9]

Some protests were filed with the government against the restructuring of the lower Puyallup. An official investigation was conducted by Captain Thomas W. Symons of the Army Corps of Engineers. He reported that the filling and location and construction of the dam were "beneficial to the interests of navigation" and recommended official approval of what had been done. It was given.[10]

On December 10, 1892, the Tacoma Land Company, the St. Paul & Tacoma, the Pacific Naphtha Launch Company, and the Commencement Bay Land Company filed a map and deeds with the Pierce County auditor showing the changes they had made.[11]

Western red cedar was used extensively for siding and shingles (from West Coast Lumberman)

"We Live Here, Too"

THE ST. PAUL & TACOMA people plunged unhesitatingly into the political and social life of the community.

When Washington became a state in 1889, a few months after Mill A began cutting, George Browne was nominated by the Republicans for the state house of representatives. He was elected by a handsome margin and at once found himself involved in the election of the new state's United States senators, the legislature then being the agency that chose members of the upper house.

Browne's position was a bit awkward. His friend and business associate, Chauncey Griggs, was a candidate for the Senate—but Griggs was a Democrat, a life-long Democrat. Browne's conflict of interest could have been more acute. As things stood, his vote could not make a difference: no Democrat had a chance. The Republicans had won 32 out of the 33 seats in the state senate, and 62 out of the 71 seats in the house of representatives. The Democrats held only eight seats in the house and one in the senate; the Independents had the odd representative. With the Republicans outnumbering the Democrats 9½ to 1, Colonel Griggs had only the satisfaction of winning all nine votes controlled by his party. That was all he got. Browne voted Republican.[1]

The first representatives served only one year. Browne declined to seek re-election in 1890. He was suggested as a Republican candidate for mayor of Tacoma but declared himself unavailable, accepting instead appointment to the Tacoma Park Board.[2] A man of culture and sensibility, widely traveled, for some years resident in France, Browne loved parks and trees. He imported at his own expense poplars that he then planted along E Street (Fawcett Avenue), which had been clear-cut when it was graded. The softening foliage of the fast-growing trees was so popular that the city council adopted an ordinance stipulating that if Tacomans beautified parking strips, the city would pay the original cost if the trees survived three years. The green of many North End streets is a heritage of that ordinance.

As park commissioner, Browne took over the selection of trees and shrubs for Wright Park, which had been denuded of its native forest on the theory that there were too many firs and hemlocks around town anyway. Browne had enjoyed watching the children of Paris gather nuts in the Tuileries

C. E. & HATTIE KING,   WASHINGTON TERRITORY VIEWS.   TACOMA, WASH. TER.

Wright's Park had already been logged of its native conifers when George Browne began selecting exotics for its beautification (C. E. and Hattie King, photographers; Historical Photo. Coll., UW Library) and the Bois de Boulogne, so he saw to it that many walnut trees were scattered through the park. He favored rhododendrons and imported many varieties from Europe (he especially liked a Belgian strain), and as broad an array of exotic hardwood as he thought might prosper under Puget Sound's filtered sunlight.

Wright Park attended to, Browne later picked out the route for the Five-Mile Drive around Point Defiance, choosing a carriage route that would disturb none of the larger trees yet afford frequent views of Dalco Passage and the Narrows. Browne's decisions in those days contributed as much to the beauty of his adopted city as any man's. In them he was influenced by the genius of Frederick Law Olmsted, who created Central Park in New York but whose design for Tacoma had been rejected by the Tacoma Land Company.[3]

In 1892 Henry Hewitt was mentioned as a candidate for the United States Senate. The *Everett News* asked, editorially, whether if Hewitt were to run, "does anyone entertain for a moment the idea that he would fail to attain the object of his desire. His record of success is as yet without a break." But Hewitt did not choose to run.[4]

As for Griggs, his failure to win a seat in the Senate did not in the least diminish his interest in politics. In 1890 the secretary of the Central Democratic Club of Pierce County remarked that the party in Pierce was "dominated by four gentlemen who write their autographs Stuart Rice, Hugh Wallace, Henry Drum and Chauncey Griggs." Some

Democrats declared that the statement was "not only unnecessary and premature but very non-Jeffersonian"—but no one argued that it was untrue.[5] Griggs was a force to be reckoned with in political affairs in the new state.

Though basically a conservative Democrat, favoring the gold standard over silver in the great controversy that divided the party through the 1890s, Griggs was innovative in some of his political suggestions. He urged, unsuccessfully, that the city council institute a program of public works so that workers drawn to Tacoma by the great boom could support themselves while looking for private employment:

"We should give them all work till they can get themselves homes," said the Colonel.

There is no doubt of two premises upon which to build an argument on this employment question. One is that there is a prodigious amount of work going on in our city; the other—a too palpable fact, apparent even to the most casual observer—that there is an enormous floating population made up in the most part of unsuccessful seekers after employment. This element, brought in on the crest of the tidal wave of immigration, floats around the city for a few days, more or less, then floats off down the Sound and finally loses itself.

This floating population is not to any extent composed of the drones of the hives of labor, but of honest sons of toil attracted here by the reports on all sides of the phenomenal growth of the city, the gigantic lumber factories and the thousand and one industries which make Tacoma today one of the most buoyant labor markets on the continent. The reports are not in themselves misleading, but those reading them and most interested in the matter do not stop to think that others as well as themselves receive the same reports and are apt to be attracted by them.

The city should provide a hole for every peg and save this floating population to the city by giving it employment in paving and grading the streets, and other public work which will redound not only to the city's fame and glory but will repay a hundred fold the money invested in giving employment to honest labor.[6]

In 1892, the year that Hewitt was discussed as a candidate for the U.S. Senate, Griggs was mentioned as a Democratic possibility for governor. "Because of his large wealth and friendly relations with many of the largest institutions in the state," said the *Tacoma News,* a Democratic organ, "Colonel Griggs is looked upon as a strong candidate, one who could divide the John H. McGraw strength and bring victory to the party."[7] Griggs refused to run (Republican McGraw won easily), but he did accept the chairmanship of the Democratic delegation to the national convention in Chicago, where he supported the successful candidacy of former president Grover Cleveland.[8]

After Cleveland's election, Griggs agreed to make his second race for the Senate. The Republicans held an overwhelming majority in the Washington legislature, 75 seats to 28 for the Democrats and 9 for the Populists. But the Republican vote was divided between two strong candidates—the incumbent John Allen and Spokane attorney George Turner—and they detested each other. It was possible that the followers of one or the other might bolt to a moderate Democrat.

When the legislators gathered in joint session to select the state's next senator, 57 votes were needed for election. On the first ballot on January 17, 1893, Allen had 49 votes, Griggs 27, Turner 26, and Govnor Teats of Tacoma, the People's Party candidate, 9. The voting continued almost daily for seven weeks, the tide shifting only a little as various backroom deals were struck. Allen peaked with 52 votes, 5 short of election. Turner rose no higher than 28 votes; Griggs had a high of 28 also. Teats had only the Populists. After 88 ballots, Griggs released his supporters who drifted off to various Democratic dark horses, none of whom received more than a handful of votes.

After 101 ballots, when Allen had 50 votes, Turner 24, and Griggs only 8, "with the rest scattered about the field like bodies after a battle," the legislature simply gave up and went home without electing anybody. Governor McGraw reappointed Senator Allen to a six-year term. The courts held the appointment unconstitutional. For the next two years, Washington was represented by only one senator, Clark Squire.[9]

That 1893 attempt was Griggs's last bid for elective office. He was approached in 1896 but declared himself "too busy."[10] When one of the St. Paul & Tacoma officers made it to the Senate, it was not Democrat Griggs but Republican Addison Foster.

All the company men, or at least the officers down to the

rank of lieutenant, were encouraged to take political action. Percy Norton, the assistant treasurer, proved the most effective. He ran for the city council in 1895 as a reform Republican, was elected handily, and was serving a fourth term when he died in his sleep at the age of forty-four in 1902.

Norton had been encouraged by Griggs to enter local politics as the result of a controversy concerning the city government's purchase of the Tacoma Light and Water Company from Charles B. Wright. Wright, a Philadephia financier, had been looked on as Tacoma's fairy godfather. Of immigrant Irish stock, a boy who rose from rags to riches in the tradition of Horatio Alger fiction, Wright was a member of the committee of Northern Pacific officials who recommended Tacoma as the western terminus of the transcontinental. He had served several years as president of the Northern Pacific and as president of the Tacoma Land Company. His personal investment in Tacoma was considerable, and he had given substantially (and with substantial publicity) to worthy causes, largely Episcopalian. But he had his own interests to protect and by the 1890s Wright had become the symbol of what anti-Establishment figures, not just Populists but also estranged businessmen, called "the railroad crowd."

Tacoma Light and Water was a focal point of discontent. The water supply it delivered was inadequate, uncertain, and impure. Even the Chamber of Commerce was upset by Wright's explanation that the town had grown faster than

Percy D. Norton was active in reform politics in Tacoma until his sudden death at the age of forty-four. The city council built a fountain in his honor at the juncture of St. Helens and Tacoma Avenue (from Mississippi Valley Lumberman, *May 25, 1894)*

he expected. When a city council committee headed by an attorney on retainer from the Tacoma Land Company, which Wright controlled, negotiated the city's purchase of Wright's utilities at a price awkwardly above that set by the appraisers, Colonel Griggs was indignant.

Citing his six years' experience on the board of water commissioners of St. Paul, during which he played the leading role in developing that city's water system, Griggs offered to build Tacoma a utility that would provide purer and more abundant water and donate it to the city free if he could not beat Wright's price.[11] When the council refused to consider his offer, Griggs publicly opposed the bond proposition submitted to the electorate to finance purchase of the utility. It passed.

Later the maverick city attorney James Wickersham brought suit against Wright, charging fraud, and won a million-dollar settlement for Tacoma. That led to a move by the city government to forfeit payment on the bond issue. Griggs refused to support default, declaring that a debt should be honored, even if foolishly contracted. He explained his position in a letter to a friend in New York:

Every officer of the St. Paul & Tacoma Lumber Co. was entirely opposed to his proposition and worked early and late against carrying the same until the close of the election and contributed time and means to inform our people of the gigantic swindle which we believed would be perpetrated upon them. But after the confirmation . . . we have taken no part whatever in any movement in regard to invalidating. . . . In fact, we have

held that it was the duty of the people to take the medicine and pay the penalty. There is not an officer in our company who is in favor of repudiation of public or private debts.[12]

It was a principle that was soon to cause Griggs much inconvenience, but no doubts.

Griggs's responses to the utility futility was to throw the influence of St. Paul & Tacoma into a campaign to elect city officials not controlled by the Tacoma Land Company.

"The city government has really been in [Tacoma Land Company] hands," he wrote to a friend, "or I might say, such men have been elected to fill various positions who have appeared to be entirely under their control, or willing to labor for any measure that they sought to carry, and in many cases expending double the amount that should have been for the improvement or necessities required for our city, and it would have been better if a large portion of them had never been performed at all."

Griggs got together with other leaders of the business community to find candidates who represented his views of public service, Norton among them. In a letter to Henry Hewitt, he wrote, "Percy [Norton] has been nominated for Alderman on the Republican Ticket at the solicitation of the best businessmen in the second ward. We [are] endeavoring to get men who can be trusted and will work for the people's interest instead of individuals, corporations or cliques. If we succeed in getting in three or four good councilmen, I have no doubt we will bring the city out in good shape, but it will take two or three years to restore entire confidence."

During the ensuing city election, Griggs received a pained letter from Isaac Anderson, manager of the Tacoma Land Company, who complained that Griggs was being quoted as impugning the integrity of the company and himself.

"It appears you would lead up to a controversy regarding the present deplorable conditions of our city finances and correctness of the views of those largely interested in endeavoring to put the city where its credit will be untarnished and every city and property holder benefitted," Griggs replied. He reminded Anderson that during the bond campaign he had warned of what the consequences would be to the land company and to the city.

Griggs noted that he had spoken to the club members in private discussions. "If I desired to make any public charges against you I would have informed you regarding them before doing so, as I did at the time of the water election. It makes no difference to me that we differed in the past. We should now pull together as an entire body for the future interests of our city and our country. . . . Your company has much larger interests in this city than anyone else and I trust you will feel the same interest and work in harmony with the same objects that have been taken up by our business and professional men." [13]

Norton won election to the council. A few weeks later, in a letter to Jones, Griggs said, "Percy of course has his hands full, but in the main he can handle the business that he is doing and do justice to the city's interests. He will have to turn off the fellows very lively who are calling upon him in that line and give what time he can spare to the general interests of the city without being importuned by men who are looking for jobs or positions."

Norton's four terms covered a period of great political divisiveness and tumult, but no financial scandals broke while he was in office. He was never accused of putting his interest or that of his company ahead of the city. He was twice president of the city council and when he died in office, the city erected a fountain in his honor in the small triangle near the Temple Theater where St. Helens and Tacoma Avenue converge.

As demonstrated by his political correspondence, Colonel Griggs, though usually gracious and urbane, could be tart.

Chester Thorne, a wealthy New Yorker, had moved in next door to the Griggs family on Tacoma Avenue. He was a desirable neighbor and was treated as such until the morning when a servant tethered Thorne's cow on the border between their properties. She damaged a young fruit tree that Griggs had planted.

The Colonel penned an indignant letter demanding redress and the promise of no further bovine intrusions. Thorne responded with a check for five dollars, drawn to his servant's account. Griggs returned the check with a friendly note acknowledging that he had written in anger. He said he did not want money, only the assurance that Bossie in the future would be pastured at some remove from the Griggs' incipient orchard. After that the cow was taken to Wright Park to graze.

Tacoma's upper crust—the term "establishment" had not come into vogue—tried earnestly to get along with each other. They comprised an instant aristocracy, drawn from far places to the new town, distinguished from the motley by wealth, family, or talent, and sometimes by education. Nearly everyone was new in town. They brought sectional and sectarian differences: it was more fashionable to have been in the Union than the Confederate Army, socially most prestigious to be Episcopalian, Congregational, or Methodist, certainly not a disadvantage to be from New England or Philadelphia—but the greatest handicap was to be without money.

The well-to-do lived very well. The houses were large, comfortable, always with a view of salt water and snow mountains. The social center of town moved steadily northward along D and E streets, across Division, to Buckley Hill between the gulches. The residents vied with each other in the handsomeness of their carriages, the sheen of their horses. Racing on Pacific Avenue was something of a scandal, and the establishment of a race track at Fifteenth and Sprague failed to slow things down. Runaways were not infrequent. Chauncey Griggs and Addison Foster were thrown from the Griggs's carriage when a rein broke and the team bolted. Foster escaped with a sprained ankle but

Griggs landed on his back and was confined to his bed for some time, not a total inconvenience since he conducted a voluminous correspondence in longhand and often wrote in bed, and he was attended by a physician friend and neighbor.[14]

Dr. George Corydon Wagner, a native of Canada, descended from German forebears who migrated first to Holland, then to New York when it was still Dutch, and, as Loyalists (Tories) during the Revolutionary War, to Ontario. Wagner came to Tacoma in 1888. He occupied an office downtown for a year, then built a home and office at 324 North E, directly across from the Hewitt home. He became the personal physician and friend to the St. Paul & Tacoma hierarchy, courted Colonel Griggs's daughter Heartie, and on June 7, 1893, in the event of the Tacoma social season, married her.[15]

It was wonderful to be young and affluent in Tacoma in the early 1890s. The land was lovely and the city charged with energy. People took seriously the slogan coined by the eccentric promoter George Francis Train, "City of Destiny." They heard purpose in the screaming saws, saw it in the smoke plumes from the waste burners, smelled it in the sharp scent of freshly sawed lumber. They were building a metropolis. Work was rewarded not merely with wealth, it was an act of creation. And there was time for play.

Japanese lanterns were in vogue. In the spring and summer, when parties were given in gardens, the hillside above the bay glowed palely with lamps moving gently in the breeze. Sundays were for sailing on the bay—one photo shows seventy-two small boats under sail in front of the Tacoma Hotel—or taking launches to the neighboring beaches for picnics. It was exciting just to take a streetcar ride to the end of one of the lines for a picnic in forest or prairie. On great occasions there was the trip to the Mountain by train and horse. Women were taking up climbing and a young schoolteacher from Yelm, Fay Fuller, had reached the top.

The townsfolk in the 1890s formed a yacht club, a tennis club (George Browne was its first president), a bicycle club, and a fox-hunting club. In 1894 three young men working for the Balfour Guthrie Company began playing golf on a field of a suburban farm two and a half miles southwest of town. They imported niblicks and mashies, drivers and putters from Scotland, and before long had enough men interested in the pursuit to form the first golf club to be organized on the West Coast below Canada. On May 22, 1895, a group of business and professional men were sitting on the porch at the home of Charles B. Hurley, who had come to Tacoma to manage Wright's Light and Water Company a few years earlier. Among those present were Dr. Wagner and L. B. Lockwood, who was law partner to Herbert Griggs. The talk turned to golf and before the day was out the men had organized themselves as the Tacoma Golf and Country Club, with headquarters on the north side of American Lake.

Culture remained the province of the ladies, though the men did organize a Washington State Historical Society (of which Henry Hewitt eventually became president) and an Academy of Science (with C. H. Jones as a trustee), which devoted most of its time to defending "Mount Tacoma" as the correct name for the Mountain.

The first women's club of which there is record was the Tacoma Art League, which met for talks on art and displays of paintings by the members. Mrs. Griggs, who had studied painting for a year in Europe and was a talented amateur, served first as vice-president, then as president of the organization. Colonel Griggs was a trustee, and Herbert Griggs served as counsel. When the League was opened to men, Chauncey Griggs and George Browne became members. Meetings were frequently held at the Griggs's mansion.

In May of 1892, five months after the organization of the Art League, some of its members met for a social gathering at the Samuel Slaughter residence at 608 South G Street and decided to organize a second group with broader horizons than representational art. At a second meeting at the Griggs place, a constitution and by-laws were adopted. The ladies included as the purview of their club, "literature, art, drama, music, science, education and philosophy."

Martha Ann Griggs, only recently returned from a trip to Hawaii, suggested the name "Aloha" for the organization, giving it the original Polynesian meaning of "kindness, love or affection" rather than that of a salutation of greeting or farewell. "Aloha" it became and still is. Membership was limited to sixty women, among them Mrs. Griggs, her daughter Heartie, Mrs. Browne, and Mrs. Harrison G. Foster, the wife of Addison's eldest son.

In addition to clubs with a cultural orientation, Tacoma had its share of secret societies whose rites might be abstruse but whose members and their social positions were well known. Ethnic groups abounded, especially among the Germans and Scandinavians. There were also church auxiliaries, veterans organizations, and unions aplenty. Benevolent and protective associations with good intentions and peculiar names sprouted like alder in a logged-off area.

In a time before government welfare programs, the staging of benefits to raise money for worthy causes took much of the time and ingenuity of well-to-do women. Something new, something that could be charged for, was a matter for thought and research. During the fall of 1892, when the men were preoccupied with the attempt of former president Cleveland to win his old job back from President Harrison, a young society woman belonging to a group styling itself the King's Daughters came up with a novel idea for fundraising: fortune-telling.

Divination had become a fad in the East. The notoriously hard-headed Commodore Vanderbilt had left something like $100,000 in trust to promote spirit-rappings. Even non-believers professed open minds. To clinch her argument, the young woman reported that a certain Mrs. Boffler, new to Tacoma and reputed to be a clairvoyant of world rank, was willing to reveal the future for a dollar a seance and turn all proceeds over to the King's Daughters. Martha Ann Griggs agreed to be hostess for the event.

In the weeks before the party was held, Mrs. Boffler appeared uninvited at the homes of several prominent Tacomans to offer previews demonstrating her gift of second sight. According to the newspapers she "certainly does possess a most wonderful gift, for she told a great many secrets such as engagements, entanglements, flirtations and various other trifling but interesting society secrets, and just a tinge of that undercurrent of things whispered but not spoken of in society that lent a spice to her recitals." After each such visit, the people for whom she prophesied received a telephone call from the police suggesting that they count their silver and inventory their jewelry. In no case was anything missing, but the notoriety and mystery of Mrs. Boffler boosted interest in her impending performance.

Came the great night. The Tacoma elect assembled in the big house north of Old Woman's Gulch to await the arrival of the seeress. Mrs. Boffler came late, a sallow, snag-toothed crone wearing an old-fashioned gown, oddly shaped glasses, high-button shoes, and a lavender bandana, which, like the hair beneath, needed a wash. Her voice was nasal and twangy, her insistence on having things exactly to her specification was adamant. The room must be darkened. She would see but one person at a time. She must not be told who they were as they entered. Her conditions were met and the fun began.

Mrs. Boffler was better than advertised. She mentioned things that no outsider could know about the society crowd. She was alarmingly frank. One matron left in hysterics, another in tears. She told a recently married woman that she had taken unto herself a spouse "for revenue only." The men fared better. The *Daily News* reported that among the gentlemen who submitted their souls to be read were Percy Dickinson, railroad contractor; George Milliken, mining expert; Lester Lockwood, lawyer; Dr. Wagner; and Everett Griggs. Dr. Wagner was said to be so flattered by what he learned that he paid an extra installment to the King's Daughters. Dickinson proferred a five dollar tip. Everett Griggs asked for a second sitting in hopes of getting her to clarify matters on which she touched only lightly and mysteriously, but Milliken stalked out and went back to his rooms at the Union Club.

Mrs. Boffler's baffling performance became the talk of the town. The sensation was not diminished a few days later when it was discovered that the fortune teller was a fake. "Mrs. Boffler" was, in reality, the young woman who had recommended her to the Daughters. By removing three false teeth, broadening her nostrils with bits of sponge, discoloring her complexion with coffee stain, adding a wig and glasses, she had escaped recognition, even among friends. Her foresight was a matter of memory.

Those who had been hoodwinked and affronted could find some solace in the fact that the soiree, after expenses, had raised thirty-one dollars for the King's Daughters charitable efforts.[16]

The Everett Connection

Within the last six weeks, if reports in the papers are correct, Henry Hewitt Jr. of Tacoma, Wash., has decided to perform the following feats: Build a sawmill at South Bend, Grays Harbor or Puget Sound, erect a pulp and paper mill at South Bend, open two stone quarries and dig a well in Chehalis county, build a logging road in Clallam county, and a boom in Thurston County. In addition to this he has bought a few thousand acres of timber land and also went to church. People who do not know Mr. Hewitt very often misread his features. He is not cross-eyed, but he does not always look where he is going to strike.

—Undated clipping, Hewitt Papers,
Washington State Historical Society

WITH the start of production in the big mill in 1889, Henry Hewitt felt free to engage in his favorite pastime, the acquisition of more timberland and mining claims. He roamed western Washington by horse, boat, and foot in search of exploitable resources. In two years he picked up 40,000 acres of timber for St. Paul & Tacoma adjacent to their original purchase, 50,000 acres of timber for himself, mostly in Lewis County, and 3,000 acres of coal and mineral land. As a member of another syndicate he secured 16,000 acres around Port Gardner and along the Snohomish River. The last of these purchases caused concern in Tacoma as it involved the creation of a new and presumably rival industrial city: Everett.[1]

Attracted by an advertisement in an obscure weekly, Hewitt had gone north in 1890 to look again at timberland on the Snohomish. The trees were splendid, so he checked to see if there were an area on salt water where loads could be boomed for tow to the mills. That search brought him to Port Gardner, a shallow bay at the mouth of the Snohomish, scantily settled. Hiring some Indians with a canoe, Hewitt sounded the bay and found, off the delta, depths ample for the deepest keel.

On returning to Tacoma, Hewitt was asked to join a party which included two Eastern financiers—Benjamin Lombard of the Lombard Investment Company of Boston, and Harry Davis, secretary to Thomas Fletcher Oakes, now president of the Northern Pacific—who had been sent west to investigate the possibility of building a railroad to Bellingham. The group went up the Snohomish River as far as the town of Snohomish; up the Skagit to the Hamilton mining district, reputedly rich in coal and iron; and across the flats to Bellingham. They reported that not only could rails be laid without great difficulty but that the line would tap rich lumber and mineral valleys. The report gathered dust.[2]

Another scout who came west that year out of the camp of eastern capital was Charles Colby, wealthy in his own right, the son of wealth, but who, even more importantly, had a special relationship with John D. Rockefeller. The Standard Oil mogul was making more than ten million dollars a year and could not think of ways to invest it all. Colby and Colgate Hoyt, his partner in a New York brokerage house, saw Rockefeller regularly in the Fifth Avenue Baptist Church. Hoyt and Rockefeller had been acquaintances in their youth. The partners knew business, they worshipped right, and Rockefeller trusted them. When they found ventures worth investing in themselves, Rockefeller put some surplus earnings into the enterprise. Colby and Hoyt became Rockefeller's designated spenders.[3]

Though born in Boston, Colby had spent several years in Wisconsin, looking after his father's railroad interests and serving in the legislature. High among his enthusiasms was the American Steel Barge Company of Wisconsin. First established to fabricate barges for Mississippi and Great Lakes traffic, the company in the late 1880s began building whaleback steamers.

The whaleback, conceived by Angus Alexander McDougall, was designed to cruise half-submerged, so low in the water that, in theory at least, it would avoid the violence of ocean storms. Three whalebacks had been built. The *Colby Hoyt* was serving in the Great Lakes; the *Joseph L. Colby* had made it down the St. Lawrence and was working on the Atlantic Coast; and the *C. W. Wetmore* was scheduled to descend the Mississippi, carry cargo to England, and return with pipe for the Tacoma Water Company and equipment for a new barge factory to be built on the Pacific Coast.[4]

McDougall accompanied Colby to Puget Sound to look for a site for the western barge works. Colby had known Henry Hewitt in Wisconsin and called on him at Tacoma. Hewitt introduced him to officials of the Tacoma Land Company and the smelter. Colby found them surprisingly cool to the possibility that he might locate a barge works on Commencement Bay and he looked elsewhere. He had about decided to locate at Anacortes when he chanced to cross paths with Hewitt again. Hewitt mentioned the possibilities of Port Gardner, and Colby was sufficiently interested to invite Hewitt to accompany him on a trip to southeastern Alaska so they would have a chance to talk.[5]

By the time they returned, Hewitt was no longer thinking about Port Gardner as a log-booming site, but as a boom town, and Colby had expanded his shopping list to include sites for a nail factory, a pulp mill, a sawmill, a railroad, a dry dock, warehouses, and hotels. In short he was thinking of creating an industrial complex, the zenith city of Puget Sound, even the Pittsburgh of the Pacific—with the help of Rockefeller money, of course.

Hewitt, accompanied by McDougall and Davis, went back to Port Gardner for a final check. Deciding it was suitable for industrial development they went to New York to confer with Colby and Hoyt. When Hewitt returned to Puget Sound he carried authority to spend up to $800,000 Colby–Hoyt–Rockefeller money to acquire a townsite. He also bore a new name for the proposed development: Everett. He was later said to have suggested it when Colby's son Everett asked for seconds on dessert. "That boy wants only the best," said Hewitt to his host. "So do we. We should name our city Everett." [6]

The Colby–Hoyt group was not alone in appreciating the possibilities of the bay at the mouth of the Snohomish. Hewitt found that some other Tacomans—the Rucker brothers, Wyatt and Bethel; their formidable mother, Mrs. J. M. Rucker; and their friends, Mr. and Mrs. Swalwell—had taken homesteads and purchased claims along the waterfront and the river. Along with the handful of early settlers, they held most of the key land, and the Ruckers did not have farming in mind. They, too, were thinking in terms of a city. Rich in faith if not capital, they had already paid for the survey of fifty acres, filed with the county a plat for the town of Port Gardner, and set about clearing.

Land deals were second nature to Hewitt. He quickly picked up an old claim at the mouth of the river for $20,000. He learned that another desirable chunk of real estate, the homestead of a deceased Mexican War veteran, was in the not-fully-certified possession of a splendiferous old saloon-keeper, Emory C. Ferguson.

Old Ferg had drifted north from the California gold rush, bedded down beside the Snohomish where he staged a memorable Fourth of July celebration to which no one came, "not even an Indian," and out of sheer loneliness and optimism opened the Blue Eagle Saloon. Having the good fortune to be the only confessed Republican within miles when Abraham Lincoln appointed Washington Territory's first Republican governor, Ferguson collected patronage appointments by the bushel. To quote from Norman Clark's splendid profile in *Mill Town,* Old Ferg stood ready "to serve food or drink in his capacity of bartender, to hand out mail as postmaster, to dispense justice as probate judge and justice of the peace, to transact business for the county as county commissioner and county auditor, or to discuss problems of the territory as a legislator." He also dealt in real estate. When an ill-dressed, stumpy midwesterner with a slight limp and a knack for asking dumb questions stopped by at the Blue Eagle, Old Ferg was pleased to unload on him part of the Ezra Hatch estate for $4,000. [7]

The Ruckers, who knew Hewitt from Tacoma, were not to be cozened by old-shoe folksiness. Before they would part with any property they had to be assured that enough industry would be built on what they sold to raise the value of what they retained. Hewitt, on behalf of his backers, signed an agreement pledging that a sawmill capable of cutting 100,000 board feet a day would be built within eighteen months; that a railroad would connect Everett with the proposed Seattle, Lake Shore and Eastern; and that within four years there would be on the shore of Port Gardner a barge plant costing no less than $150,000 and a dry dock costing $35,000. With that promise in black and white, the Ruckers sold 434 acres—half of their holdings—for $27,000, then donated half of their remaining acres to the cause. Hewitt persuaded E. D. Smith, the first settler at nearby Lowell, to sell another key patch on the waterfront with the promise that a paper mill would be built on it.[8]

Within two months Hewitt assembled a compact 5,000 acres, enough to make city building feasible. On November 19, 1890, Hewitt, Walker Oakes, and George S. Brown, all of Tacoma, filed papers in Pierce County incorporating the Everett Land Company.[9]

Brown (not to be confused with George Browne of St. Paul & Tacoma) was a young graduate of Brown University and the Columbia Law School. He had come west two months earlier with a classmate, Francis Brownell, to open a law practice. Brown and Brownell were fraternity brothers (Alpha Delta Phi) of Charles Colby's nephew, Gardner, who worked in the Colby and Hoyt office on Wall Street in New York. Gardner recommended his friends to his uncle as true blue Alpha Delts and Phi Beta Kappas to boot. According

to Brownell, Colby "decided to have us as local counsel for the Everett enterprises, probably because he felt we would report to Gardner more fully than would strange counsel. This was desirable because he knew that Mr. Hewitt was a man of limited education who hated letter writing and paper work."[10]

Hewitt was chosen president of the Everett Land Company; C. W. Wetmore, the New York financier and barge enthusiast, vice-president and secretary; Gardner Colby, treasurer; and young Brown, assistant secretary in charge of getting letters written. The filing of the incorporation papers triggered alarm in Tacoma. It was noted that Walker Oakes was the son of the NP president, that Colby and Hoyt were both on the executive committee of the NP board of directors, and that Hewitt was vice-president of the company with the largest payroll in Tacoma. The fear was that the Northern Pacific, having skimmed the cream from its land holdings in the vicinity of its terminus, was launching an alternate terminus, complete with a land development promotion, farther north.

Hewitt told Griggs in confidence of the plan for an industrial city on Port Gardner. Griggs expressed concern that Hewitt's involvement would create hostility in Tacoma for the company and its officers.[11] Hewitt argued that "the multiplication of thriving Sound cities will benefit Tacoma, that Everett will be to Tacoma like Baltimore to New York, or Milwaukee to Chicago, and as in the case of the Atlantic seaboard, all the Sound towns will vie with, not rival, each other and in time all must become practically suburbs, tributary to, but drawing support and prosperity from the greatest maritime mart on the west coast, Tacoma."[12] In public, Hewitt spoke only of lumber possibilities in the Everett area.

Rumor turned to boom in the spring of 1891 when Hewitt returned from a two-month visit in the East. On behalf of his associates he filed in rapid succession papers of incorporation for several new companies. There was the Everett Timber and Investment Company, incorporated at $350,000 with Hewitt as president, and Walker Oakes, George Brown, Colby, Hoyt, Wetmore, and William Henry Hewitt (Henry's son) as trustees. There was the Puget Sound Wire Nail and Steel Company, capitalized at $200,000. And there was the Puget Sound Pulp and Paper Company, with $400,000 capital stock, Hewitt as president, Lewis D. Armstrong as vice-president, Wetmore as secretary, and Gardner Colby as treasurer. At the same time a group headed by Emory Ferguson incorporated the Snohomish and Port Gardner Electric Railway Company at $1,000,000. The Snohomish, Skykomish, and Spokane Railway and Transportation Company, which had originally been formed to run a short line from Snohomish to Lowell, filed supplementary articles to enable it to start at Everett.[13]

With their secret out, ending all chance to pick up more land at low prices, the Everett Land Company commenced clearing its townsite. They cut the great trees, piled slash and undergrowth around the stumps, and set fire to the heap. You could find Everett with your eyes shut, just following the smell of smoke. The rush was on. So many came to get in on the ground floor that John Rogers, the local undertaker, boasted that his coffins were more in demand for bunks than for burials.[14]

Outlying additions sold out before lots in the heart of town were put on the market. When at last the Land Company opened the doors of its frame headquarters to the importunate customers, demand had been built so high that the property Hewitt bought from Old Ferg for $4,000 and sold to his brother-in-law, Jones, for an unstated price, and bought back from Jones on behalf of the Land Company for $128,000, was parcelled out in pieces for $500,000. There was something in it for everybody.[15]

The streets were named for the men who were building the town: Rucker, Colby, McDougall, Rockefeller, Norton. When it was realized that they had used up the best streets without naming one for the father of Everett, they ran a 100-foot-wide thoroughfare from Port Gardner Bay to the Swalwell addition: Hewitt Avenue.[16]

The editor of the *Snohomish Eye,* not a friendly witness since he spoke for a rival town, visited Everett in December of 1891 and testified:

It would convey a false impression to the outside world if somebody should bring a kodak and photograph the thing just as she looks. There are ten stumps to one house, and mud everywhere. The avenue in the distance looks as much out of place as a plug hat on a Silver Creek miner. These remarks apply to the

Everett in 1892, and the Puget Sound Wire Nail and Steel Company (from N. Clark, Mill Town; *Everett Community College Library)*

part of Everett next to Swalwell's Landing on the inside of the Peninsula. Over on the other side, next to the Sound, there is nothing but prospect. The approaching visitor sees off to his right the wide waters of the Sound and the delta of the Snohomish River. Straight ahead is more water. To the left is the townsite of Everett proper, distinguished only by its slope to the water and its gorgeous possibilities. . . . Rain doesn't count at Everett, and the residents were working as hard as Seventh Day Adventists on Sunday. There is every indication that a town will some day occupy the place. People are actually building houses there with cupolas and bulging windows on them, and act as if they intended to stay.[17]

Hewitt founded a bank in partnership with Benjamin Lombard, the Boston financier, and joined Jim Bell in building a sawmill. They put the mill on the tideflats, half a mile from shore. "The people thought we had gone completely daffy," Hewitt said later. "They nicknamed the mill The Light House. But we had a good reason for putting it way out there in the water as we had to have a harbor and this was a good way to start it. I had already built the St. Paul & Tacoma Lumber Company on similar tideflats and we found that it did not take very long to build up land from the refuse of the mill."[18]

Hotels sprouted among the stumps, some of them handsome, some even without beds. Saloons flourished; brothels, too. The nail factory was quickly completed and began turning out triangular nails, many of which were driven into the clapboards of flimsy frame houses thrown together for the assembling work force. The pulp and paper works took shape at Lowell. Equipment was installed for soda pulp production of fifteen tons a day, and a two-machine paper mill of similar capacity.[19]

One hundred and fifty men were at work putting up the barge plant. The Pacific Barge Company had been incorporated in October for $600,000, with Hoyt, Wetmore, McDougall, Brown, and Brownell among the trustees. Hewitt and Walker Oakes were to have been members of the board but were in the East when the papers were filed. The company was intended to manufacture both whaleback steamers, of the cargo type already built in Wisconsin, and passenger ships that would have cabin space "raised from the hull on mighty stilts reaching far up beyond possibility of attack by the waves."

The arrival of the *C. W. Wetmore* on her maiden voyage to the Pacific was awaited eagerly in Everett. Her design was expected to create a complete revolution in the construction of sea-going merchant ships and "do for American ocean commerce what British tramp steamers have done for the United Kingdom." Already she had left Superior, Wisconsin, where she was constructed. She had descended the Mississippi and crossed the Atlantic to Liverpool with 87,000 bushels of wheat, and returned to Philadelphia to pick up machinery for the barge works, the paper mill, and the nail plant, as well as water pipe for Tacoma.

When the *Wetmore* departed Philadelphia for Puget Sound with a December 1 estimated time of arrival, the Superior *Leader* apotheosized her creation and mission thus: "A picture which hardly falls below the quality of sublimity is that of a few landlubbers who own the American Steel Barge Works building in a little town of 25,000 inhabitants, 1,500 miles from tide water, a strange craft of solid steel to ride under instead of over the waves around the Western Hemisphere, through cyclones and tornadoes, sure, sound and safe." That hardly described the *Wetmore's* voyage.

The entire population of Everett gathered on the wharf in a storm to welcome the *Wetmore* on December 1. Captain McDougall, her designer, was present, a weathered old salt in tarpaulin hat, sea boots, and a rubber overcoat that reached to the heels. The whaleback did not reward her creator by coming round Skagit Head that day, nor for the next three weeks. When she did arrive she brought a tale of woe.

As the strange craft came round Cape Horn the water mains bound for Tacoma shifted, crushing two crewmen. As the 265-foot ship wallowed up the coast of Chile enough of the coal, bunkered aft, had been burned to cause her stern to rise so far out of the water that the propeller was damaged and the rudder loosened. She limped on, but off Northern California lost the rudder. Vibration of the screw became so violent that the engines were shut down and storm sails raised. The *Wetmore* somehow came within sight of the mouth of the Columbia River, but her captain did not dare risk the dangerous bar under sail.

The British steamship *Zambesi* took the *Wetmore* under tow and tried to bring her into the river. The hawser parted and the whaleback was almost lost in the breakers before another line could be taken aboard. Captain George Wood, one of the most respected and fearless of Columbia bar pilots, left the *Zambesi* and boarded the *Wetmore.* Under his guidance she made it under tow into Astoria. The rudder was repaired, but the *Zambesi* was awarded $50,000 for her salvage services by an Admiralty court.

The whaleback reached Everett three weeks late. Her crew struck, demanding that someone be hired to make their bunks. Nor were her troubles over. On subsequent voyages she sprang leaks, rammed wharves, lost power, or simply got lost. Within a year she was irretrievably grounded on North Spit near Coos Bay. This was not enough to disenchant Everett's developers. The barge plant was completed and work began on a new and larger version of the whaleback.[20]

Everett's greatest hopes lay in the mountains behind the town. From them the optimists expected precious metals, a railroad, or both. James J. Hill was pushing his Great Northern westward. There was every reason to believe it would be more efficient than the Northern Pacific and that its terminus would reach Puget Sound north of Tacoma. Why not at the Magic City, which was materializing out of the mud at the mouth of the Snohomish? Even if Hill hit water somewhere other than Everett, there remained the Northern Pacific. Many of its officials had invested in the Everett Land Company and associated enterprises. To compete with the Great Northern, the Northern Pacific would need to shorten its route through Washington. A logical solution would be a line through Snoqualmie Pass. Surely the Northern Pacific would not choose Seattle, Tacoma's great rival, as terminus for such a shortcut. Would they not, instead, come to Everett, thus outflanking Seattle north as well as south? For the true believer, the only question was which transcontinental Everett would get.

In February of 1892, Jim Hill, having completed arrangements in London and New York that assured him the money needed to complete construction of his main line, visited Puget Sound on an inspection trip. Learning on the morning of Wednesday, February 17, that Hill would be in Everett that evening, Henry Hewitt arranged a banquet at the Bay View House. For such short notice, it was splendid. The hall was decorated with ferns and mosses. Cut flowers and potted plants stood on tables arranged to form a hollow square. The menu included fish, fowl, beef, venison, and bear, assorted still wines (including a Chateau la Rose, 1878), and much champagne. All 102 of those whom Hewitt had invited were at the hotel before six P.M., all but the guest of honor.

Business had taken Hill to both Bellingham and Seattle before Everett. It was 8:30 when he arrived, a heavy-shouldered, full-bearded man in dark clothing, a large western homburg set square on his balding head. Accompanied by Hewitt he strode wordlessly through the townsfolk gathered to look on destiny incarnate. Hill was well aware that he carried with him the hopes of every community on the sound, that all waited his word, *the word,* on where the Great Northern would terminate. He favored Seattle, but would go there only if the city met his conditions as to space and right-of-way. Meanwhile he was keeping his options open, his mouth closed. When called on to speak, he uttered generalities; when pressed, he gave hints but no promises.

The Father of Everett and the Empire Builder entered the banquet hall to applause. Eight-thirty was mighty late for dinner in the West. Everyone was eager for food and news. The first course, soup and a fine white wine, disappeared with astonishing speed. As the plates were being cleared, Hewitt arose. Acknowledging the lateness of the hour, he introduced the visitor to his left, the only man to build a transcontinental without the assistance of the federal government. He noted that Hill had told reporters earlier in the day that he favored a fresh water harbor. Hewitt promised that Everett would supply his longing "till you can't rest." It was an offer that must have puzzled those unfamiliar with Hewitt's dream of placing a gigantic floodgate at the mouth of the Snohomish to hold water in except at high tide.

Hill avoided giving the assemblage *the word.* Instead he teased them with a comment on the multitude of self-styled cities of destiny to be found on Puget Sound. Why, he'd seen only four stumps without a town's name attached. He cautioned of the folly of expecting something for nothing; lectured on the relationship of grade to economy on long-haul railroads; praised timber as the region's greatest resource; acknowledged that Puget Sound was closer than San

Francisco to the Orient. But the nearest he came to promising them a railroad was to say that the Great Northern was a rake and that many of its prongs would touch salt water. Then he smiled mechanically, sat down, and listened with half an ear as assorted judges, mayors, legislators, and commissioners elaborated on his virtues and those of Everett.

Losing interest in the speakers and the food, Hill engaged in an animated discussion with Hewitt. The other diners watched the pair at the head table as they drew plans on the tablecloth, diagrammed the waterfront with cutlery and rolls, crossed continent and ocean with sweeping gesture. To one observer it seemed that "on the whole they appeared to agree with each other, but occasionally there would be a difference of opinion, when Mr. Hewitt would become demonstrative, gesticulating freely and President Hill would subside."

After the coffee and *kirschwasser,* while champagne still flowed, Hewitt called on Old Ferg, who assured Hill that Everett would give him whatever he needed; on young George Brown, who said that if the Great Northern were a rake, surely Everett was the point at which the handle should be attached; on State Representative Frater, who said he always opposed anti-railroad legislation (this drew applause from Hill), and on State Representative Frame, who avoided commitment and got no handclapping.

It was getting late. Hill had looked several times at his pocket watch. Hewitt rose to say he had an announcement of importance. He had just received word from the East. Colby and Hoyt assured him that money was available for a railroad into the mining district behind the town. It was to be hoped that such a line would not be necessary, that the Great Northern would bring the ore and Everett would not have to lay a rival track, but if they had to, they could. He looked to see if Hill wished to respond. Indeed he did.

Still ducking a specific promise, Hill said that in railroading there were two types of terminus. There was the actual, designated end of the line, and there was the commercial terminus, the place that carried on the greatest business. Without exactly saying so, he left the impression that such, at least, would be Everett's role. But, he added, a great deal of land would be required to build proper facilities for any type of terminus. He looked at his watch. One o'clock. Hill thanked them for welcoming the Great Northern as a neighbor and left to thunderous applause, his destination Seattle.[21]

Hewitt's threat of an independent line into the mountains was no idle bluff. Colby and Hoyt had indeed been bitten by the gold bug. The center of infection was a plateau of about fifty acres on a fork of the Stillaguamish, high in the Cascades, some sixty miles northeast of Everett. Mountains infolded the plateau on three sides and ravines radiated from its central ridge. "What a spot!" wrote an early visitor. "It is unique in beauty, sublime in grandeur, thousands of feet above the sea."[22] But the features that impressed Joseph Pearsall, a prospector from California who reached the valley in 1889, were the patches of rock laid bare by snowslides. They exposed glinting streaks of ore.

Pearsall, a solitary man and a reader, hooked on the

works of Alexander Dumas, took a claim and named it Monte Cristo. The Wellman brothers, next on the scene, staked out claims they called Pride of the Mountain and Pride of the Forest. Other fortune hunters followed. They found gold in small quantities, enough to keep a jackass prospector in grub, but the basic metals proved to be silver and lead. These were not to be panned or picked up where they lay. Getting them out required an industry. This site was remote. Hard-rock mining took money, so development was not instant.[23]

In 1891 there arrived in Seattle from Denver a free-wheeling promoter named Judge Hiram Gilman Bond. Unlike many in the West who bore the sobriquet "Judge," Bond was an attorney and had served on the bench. Finding the work of a federal judge in Virginia boring, he built an elevated railroad in Brooklyn. It made him a fortune, which he promptly lost speculating in patents. He got rich again in coal and iron in Alabama, bought a ranch in Colorado, plunged on mining claims, sold his one-sixth share in the Virginus Mine just before a shute was discovered that yielded $27,000,000, bought 136,000 acres of land in New Mexico that proved to be not only desert but also not the possession of the people who sold it to him, and moved on to Puget Sound, still questing.

At the age of fifty-three, Judge Bond was looking for targets of opportunity. Monte Cristo seemed opportune. Bond organized a syndicate that included Leigh Hunt, the owner of the Seattle *Post-Intelligencer,* and Stephen Blewett, an old-timer who left his name on a Cascade Pass. Then Bond bought out Pearsall and the Wellmans.

Bond and his two sons, Louis and Marshall, both recent graduates of Yale, set up camp at Monte Cristo, mapped outcroppings and took extensive samplings. These indicated that the most promising vein crossed through the mountain from Monte Cristo Creek on one side to Pride of the Mountains Creek on the other. Short tunnels, drilled from 25 to 50 feet into the rock at various altitudes, gave promise that commercial-grade lead-silver ores extended through the mountain.

Leaving his sons to continue the samplings, Judge Bond went to Seattle to consult with his backers. Visions of a mining district equal to the Coeur d'Alene in Idaho, Leadville in Colorado, even to Butte's "Richest Hill on Earth," rose before them. But exploitation of their claims would require far more capital than they controlled. The Everett Land Company, with its eastern connections, came to mind. Everett was the tidewater town nearest Monte Cristo. Development of the mine and the building of a railroad and a smelter would do more to boom Everett than would barge works, nail plant, and pulp mill combined. Here indeed was the possibility of a new Pittsburgh. Bond called on Henry Hewitt, who sent him on to Colby and Hoyt at 36 Wall Street, who talked to John D. Rockefeller about this new and promising opportunity. Rockefeller gave his designated spenders their head.

So it was that one day in 1892 when Marshall Bond was

leading a pack train down the mountain to Everett, he met his father on the way in.

"No use coming back," said the Judge. "I've sold the mine."

Marshall loved the high Cascades, loved the rugged life of the mining camp, felt possessive about the Monte Cristo. He protested, but Judge Bond was pleased with the deal. "I sold it to John D. Rockefeller for $800,000 and made a good profit," he explained. "Operating expenses are high and I have a hunch it isn't as good as it looks."[24]

Judge Bond was almost alone in his doubts. The state geologist after a quick survey claimed the area had enough ore to keep five or six smelters busy. "The people of Washington have not the slightest conception of the great mineral resources of the state," he said flatly. "With such mineral wealth mining is bound to become one of the chief industries of Washington."[25]

The usually cautious, conservative Thomas Burke of Seattle advised a wealthy client that "the mines in the Silver Creek and Monte Cristo district, on the west side of the Cascade Mountains, promise to be enormously rich, equalling the best discovered in Montana or Colorado. There comes from the mines but one report from all classes of men, the miner, the expert, the capitalist, all say the mines are unquestionably rich. The next move will be to secure transportation from the mines to the seaboard."[26]

Colby and Hoyt dispatched two mining engineers to Monte Cristo to make sure. The experts reported that actual development work was limited, but in view of the fact that the vein of ore was exposed on each side of the mountain they considered the information available to be adequate. They estimated from the values of the ore taken in the shot test tunnels that, if as expected the vein was constant, there would be a huge tonnage of silver-lead ore, enough to repay many times over the cost of laying rails fifty miles up the mountain side from salt water, building a smelter and a concentrator, and installing mining equipment.

The chief engineer who made the enthusiastic report had long been associated with Colby and Hoyt. He was their adviser when they brought in a very successful iron mine in Wisconsin and another in Cuba. Although he had no experience in lead-silver mines, his specialty being iron, he was generally regarded as one of the leading mining experts of the day. Without taking the precaution of extending the test tunnels deeper to see if values held, Colby and Hoyt made the plunge.[27]

Less than a month after Jim Hill's visit to Everett and well before it was known where the Great Northern's route would lead, Colby and Hoyt informed Hewitt that $1,800,000 was available to build a railroad to Monte Cristo. The Everett and Monte Cristo was incorporated on March 12, 1892. Work on the right-of-way began the next day.

Newspapers speculated that the new line would put pressure on Hill to choose Everett as terminus since, if he were to go to Seattle instead, the Monte Cristo line could be

turned over to the Northern Pacific.[28] With Everett certain to be the outlet for the Monte Cristo ore, the Puget Sound Reduction Works, which had considered building a smelter in Seattle, accepted a donation of eighty acres from the Land Company and came to Everett. Everything seemed to be falling neatly into place.[29]

Walking the board sidewalk along Hewitt Avenue in the closing days of 1892, watching the dredgers he had hired at work in the harbor, with the smelter rising in East Everett and plumes of smoke coming from the stacks of the nail factory, the pulp mill, and the sawmill, with carpenters hammering away on houses for a community of six thousand on a shore where three years before there had not been twenty people Henry Hewitt, president of the Everett Land Company, president of the Everett National Bank, president of the Everett Water Company, president of the Everett Telegraph and Telephone Company, president of the Everett Timber and Investment Company, president of the Puget Sound Wire, Nail, and Steel Company, trustee of almost every other company in town, associate of John D. Rockefeller, mentioned in the press as a possible senator or governor, must have felt the pride of a creator. All this he had conjured up out of foresight, salesmanship, brass, and energy, and with precious little outlay of his own cash. He had put up $48,000 of his own money, plus the salaries

from his various Everett enterprises and had signed notes for $150,000. That was little compared to the millions of eastern dollars he had lured west. He had been the catalyst, the activating agent that brought into fruitful ferment eastern capital and Puget Sound resources. Here around him was the city he had conceived, named, and nurtured, which he controlled so fully that he had been able to persuade the townsfolk to enter no bids against the Great Northern agent when Section Sixteen, the square mile of the township whose sale was to support the local school system, came up for auction. Hill's man paid only $76 an acre for 200 acres. The rest of the section, purchased by the Everett Land Company, brought more than $400 an acre.[30]

The sale of school lands to Hill assured Everett of a place on the Great Northern line. Though Seattle finally met Hill's demands for right-of-way and terminus property and was designated terminus, Hewitt did not accept that decision as irrevocable. He was still paying court to the railroad king. He ordered a plate made of pure Monte Cristo silver to be inscribed with an invitation to Hill to visit again "the city of smokestacks" and see its possibilities.[31]

Jim Hill did not come to Everett. He did not even witness the driving of the last spike on the Great Northern mainline at a place called Scenic on January 6, 1893. A pair of division superintendents got the connection made without

Monte Cristo from Barlow Pass area with the former Everett & Monte Cristo Railroad track in the foreground (*Juleen, photographer; Historical Photo. Coll., UW Library*)

benefit of champagne. Nor did Hill come west when the first transcontinental rolled over the Great Northern route, though he had promised to bring capitalists representing "thousands of millions of dollars" out to the end of the line to let them see for themselves the resources awaiting exploitation. "Business conditions" kept the Empire Builder in St. Paul.[32]

Business was suddenly bad at both ends of the line and in between. Business was bad everywhere. The long period of railroad expansion, which opened new lands in the United States, Canada, the Argentine, South Africa, Australia, even Russia, had created a glut in basic commodities such as wheat and lumber. The first sharp shock to the world economy was the near failure of Baring Brothers, the great British banking house, which found itself too heavily involved in Argentine enterprises. Bankers became cautious. Credit dried up. Doubt replaced confidence in the business community. Runs started on banks whose greatest resource was the public's belief that land values would continue to rise. Doubt became fear, fear bred the Panic of 1893.

Hardest hit were frontier boom towns like Everett and Tacoma, new cities with shallow roots, a floating populace drawn by the promise of easy monies, cities whose industries were new and undiversified. Everett boasted five banks on January 1, 1893; two remained on December 31. The bank failures swept away the savings of small depositors, city and county funds, and the money of eastern investors. Payments stopped on the land that working folk were buy-ing from the Everett Land Company. The nail plant shut down, the pulp mill shut down, the saw mill shut down. Work was halted on the whaleback *City of Everett*. The Northern Pacific went into the hands of receivers. The Great Northern suspended construction of all feeder lines.[33]

Hewitt's instinct called for cutting all possible expense, closing the plants, hunkering down and waiting out the storm. When good times returned the factories would be there, the timber would have added growth. No use throwing good money after bad.

Colby and Hoyt disagreed. Their proposal was to put in more money, preferably Rockefeller money. They proposed to issue bonds for $1,500,000 paying 8 percent interest over five years, secured by a mortgage on nearly all of the Everett Land Company holdings. Hewitt would not go along with the idea. Colby and Hoyt had bonds printed and signed by the vice-president. The dispute led to threats of lawsuits and bankruptcy proceedings from the eastern investors. Hewitt resigned as president of the Land Company and Colby took his place.[34]

The utility companies Hewitt headed were all losing money. He offered the light company, the water company, and the street railway to the City of Everett at less than the cost of construction. The voters turned down a proposed bond issue for the purchase.

When the Land Company held its annual meeting in May of 1894, Hewitt was not elected to the board of trustees.

As the Panic hardened into Depression, Everett's great

hope was the Monte Cristo mine. A flow of silver and lead ore could revitalize the local economy. But the riches of Monte Cristo were a mirage. Even the surface ore proved less rich than the samples had indicated. Ore values shrank as the shafts were pushed deeper into the mountain. The veins were indeed regular and permanent but the amounts of metal varied. The average value of Monte Cristo ore was .6 ounces of gold and 7 ounces of silver per ton. Lead was more abundant but was worth only about $3 a ton. It did not pay to operate the 200-ton Monte Cristo condenser. The smelter stood idle.[35]

When the Everett Land Company was unable to meet the interest payments on its bonds, Rockefeller became interested. He held most of the bonds. He had never seen Everett, or the Monte Cristo. He had been content to invest there on the advice of Colby and Hoyt. His theory was that they were intelligent businessmen, putting their own money into projects which they would examine closely since they could less afford losses than he. He was a minority stockholder in each enterprise, but when trouble came it was to Rockefeller, and the wealth others considered unlimited, that all looked for salvation. He had been drawn much deeper into the various offshoots of Everett than he had anticipated.

Rockefeller wanted out, but to get out first he had to get farther in. He bought controlling interest in several of the Everett companies, then assigned Frederick T. Gates, a lank, long-nosed former Baptist minister from Minneapolis who had become his favorite adviser, to see what could be salvaged on Puget Sound. Gates quickly forced Colby and Hoyt into the background. They were replaced by a triumvirate that consisted of Francis Brownell, of Brown and Brownell; William C. Butler, a mining engineer (and brother of Columbia University's famed Nicholas Murray Butler); and J. B. Crocker, a capitalist who had been one of Gates's parishioners when the latter was a minister.

Crocker was sent by Gates to look into every aspect of the operations around Everett. His reports have not survived, but Gates's side of the correspondence indicates that Crocker believed everybody around Everett had sought to draw Rockefeller in so deeply that he would have to keep the enterprises afloat. Gates assured Crocker that "It is quite certain that Mr. Rockefeller is not, at the present moment, prepared to put new money into the enterprise."

Gates came west for a short visit to inspect the Everett and Monte Cristo railroad and mine. The scenery in the Cascades delighted him. ("These snowcapped mountains were hardly less imposing and awe-inspiring than the Alps around the Jungfrau and Matterhorn. . . . I used to ride up on the cowcatcher and be regaled at dinner with delicious trout caught in the stream.") The balance sheets were not so delectable. He shut down the mine and unloaded the Everett and Monte Cristo on the Northern Pacific. The smelter kept going for a time on ores from the Kootenays and Coeur d'Alenes but finally was sold to the Guggenheims, who had it dismantled brick by brick. The nail mill was sold to the American Steel and Wire Company. The barge works shut down after launching only one whaleback. The pump mill was sold to a local operator.[36]

During the reorganization, when everyone was pointing fingers of blame at everyone else and recriminations were at their bitterest, Hewitt was threatened with charges of misappropriation of funds and summoned back to New York by Rockefeller. He later said that he was being set up as the fall guy, that while he had invested $68,000 of his own money in the enterprises in addition to his salaries, he owed the Rockefeller people $150,000 and could have been frozen out.

Before heading east, Hewitt took the precaution of organizing the Hewitt Land Company to which he deeded most of his timberland and mineral holdings, then distributed stock in the company among his creditors. He was in New York for two weeks, he said later, some of the time face-to-face with the fifty-five-year-old Rockefeller, then the most powerful capitalist in America.

A reporter who interviewed Hewitt fourteen years later for a profile in the *Tacoma Ledger* wrote:

When he returned to Tacoma he brought with him the honor of having been the first and only man ever to get the best of John D. Rockefeller. All charges of misappropriation filed against Hewitt were withdrawn and he was given greater managerial powers than he ever had before. . . . Rockefeller assumed all of Hewitt's outstanding obligations in exchange for paper mill and factory stock. Rockefeller also gave Hewitt considerable land near Everett, several mortgages and $14,000 in cash, stipulating that Hewitt remain in charge of the works at Everett and continue managing the two banks.[37]

Archivists at the Rockefeller Archive Center at the Rockefeller University in New York say no interview with Hewitt is specifically mentioned in their records, nor is any summary or an agreement reached between Hewitt and Rockefeller extant.[38] A telegram from Gates to Crocker says that on February 2, 1895, Hewitt sold his interest in the Land Company to Rockefeller. Letters indicate the price was something less than $4,000. In a letter written on August 14, 1895, Gates commented that he supposed "Hewitt, following the 'Leech and suck spirit,' preferred to have the capitalists who were behind him at all times committed to the greatest possible expenditures in Everett."[39]

Hewitt declared later that Rockefeller not only had left him in charge of the surviving banks in Everett but had authorized him to draw up to $20,000 against Rockefeller's account to keep them open. He had seen them through the worst of crises without using Rockefeller money but sold out when the situation bettered "because he had still greater enterprises elsewhere demanding immediate attention."[40]

Though he withdrew entirely from his Everett enterprises, Hewitt left behind the foundation of an industrial city. In a letter to a St. Paul banker, Colonel Griggs said,

"Our Mr. Hewitt has settled his matters up in full in New York relating to his Everett deal, which will place him in better shape, and notwithstanding the losses he has met with, he will come out with over a million dollars which in time will net him a great deal more money in my judgment."[41] If so, Hewitt may indeed have been, as he claimed, the only man to best Rockefeller in a confrontation when the master was at the height of his power.

No logging photographer could resist arranging the crew in this pose (from *author's collection*)

Crash and Crisis

THE PANIC that destroyed Hewitt's early plans for Everett ended Tacoma's dream of becoming the metropolis of Puget Sound. From a claimed population of 54,000 when the depression hit in 1893, the city fell to an official 37,000 in the 1900 census. One out of three residents fled the city. The collapse would have been greater had it not been for the stabilizing influence of the St. Paul & Tacoma Lumber Company and its officers, especially Colonel Griggs.

At the 1892 annual meeting of company stockholders, Griggs and Hewitt each voted 2,590 shares of the 10,440 issued. Jones had 1,000 shares and Foster 875.[1] The balance in voting strength between the Wisconsin and Minnesota pairs who formed the company remained approximately equal, as it had from the day of incorporation. But with Hewitt concentrating on his Everett venture, Jones spending much of the year in Menominee and Menasha looking after the Hewitt–Jones enterprises, and Foster spending much of the year in St. Paul in connection with Griggs–Foster enterprises, Griggs made most of the administrative decisions and set policy for the lumber company, of which he was president and chairman of the board.

Griggs was a methodical man, genial in social situations, fond of dancing and cards, but given to thinking things out in solitude and putting his ideas in writing, especially as letters to close friends and his children. It was his custom to read for an hour before going to sleep and for half an hour after being awakened at seven. He breakfasted—usually on steak, eggs, toast, mush, and coffee—at eight, and left for the office at eight-thirty, usually driving his own carriage, accompanied by Hewitt or Foster if they happened to be in town. His favorite route took him along Tacoma Avenue to St. Helens, down to C Street (Broadway), left to Pacific Avenue, right out to South Twenty-third, then right again to Holgate. That put him at his rolltop oak desk by nine.

He spent much of the morning on correspondence and studying production records. If there were to be conferences with subordinates they were scheduled for ten. He usually had lunch at the Tacoma Hotel, in the Stone Room if it involved business; when he didn't go to the hotel he favored one of the sawdust-on-tile oyster houses along Pacific Avenue.

In the afternoon Griggs visited the mill or the offices of some other business he was involved with, then checked in

Chauncey Griggs at the time of crisis for the lumber company and the city of Tacoma (Minnesota Historical Society)

at his own office before going to the Union Club, where the movers and shakers of the community talked business and politics, or played dominoes, whist, and occasionally poker. He was home at seven for dinner. There were usually callers during the evening and again much talk of politics and business. Could Grover Cleveland be returned to the presidency? Could Tacoma's government be purified? How serious was the business slump?

The decline—actually a slow-down in expansion—had started in 1891. By 1892 many mills in the Pacific Northwest were reducing their cut and some had closed, but not the big mill on the boot. With its superior equipment, its own forest and logging railroad, its yards in eastern Washington and California, its cargo trade, and its experienced management, St. Paul & Tacoma was in the best position of any Puget Sound operation to sail against the drift of the economy.

Early in 1892, Griggs visited California where he found the trade "in a very demoralized condition, not only in northern but southern California as well, with lumber being sold at about the actual cost of manufacturing and handling." He was able to turn the slump to company advantage by obtaining majority interest in the Kirchoff–Cuznor Lumber Company of Los Angeles, one of the largest distributors in southern California.[2]

Local trade in western Washington Griggs described as "very light." St. Paul & Tacoma had to deal with desperate rate cutting by small operators who had no cargo or rail outlets, but eastern sales held up remarkably through the year. Griggs had sent a crack salesman, Emory White, to the Atlantic seaboard with what the *West Coast Lumberman* described as "1 handkerchief, 2 socks, 1 collar, 1 shirt, 1 pipe, 4 boxes of tobacco and a nice line of spring styles and cuts of Fir." White sent back orders for almost eight million board feet, most of it "in better grades of lumber, better than common, with a large proportion for use in general building." In May there came orders from Baltimore for ten carloads of fir for shipbuilding and heavy construction, one calling for pieces in such special sizes as 18-by-24 inches by 68 feet; 24-by-30 inches by 50 feet; 20-by-24 inches by 58 feet, and a few even larger.[3]

When Colonel Griggs went to Chicago in June as the chairman of the Washington delegation to the Democratic convention,[4] he found time to make some sales of fir for railroad car decking and, in a new but minor market, several carloads of spruce for piano sounding boards.[5]

Though prices continued to sag, business through the year was brisk enough for Griggs to keep one side of the two-bandsaw mill operating nights as well as days. The logging railroad was extended four miles toward Wilkeson, tapping a fine tract of timber.[6] In December, Griggs told a reporter that St. Paul & Tacoma was planning to build the largest planing mill on the West Coast in the spring. It would be an all brick structure equipped with the most advanced machinery and supported by an immense lumberyard where the lumber could be sorted and air-dried pending shipment. Also under consideration was a new shingle mill with a minimum capacity of 500,000 shingles a day, and the company was committed to building the companion to Mill A within three years.[7]

For a few months in 1893 it really seemed possible that the St. Paul & Tacoma was big enough and efficient enough to prosper during a national slump. The company in January won acceptance from the Shingle Manufacturers and Dealers Association of Oregon and Washington of a pledge to maintain prices, though the agreement was less honored in observance than in the breach.[8] In March, the big mill was working on a special order of one hundred carloads of fir to be used in building an ore dock for the Duluth, Missabe and Northern Railroad. Other cars were being freighted with sills for Brainerd, Minnesota, car siding for Detroit, car roofing for Omaha, and car decking for Baltimore. At the ocean wharf four ships were loading for San Pedro, another for San Francisco. Totals for the first six months of 1893 showed twice as many ships loading cargoes as in the same period of 1892, but individual orders were smaller and profits were falling. The value of cargoes shipped was up 10 percent. In June, no new orders came in, the first such month in the company's history.[9]

In July, Griggs found it necessary to lay off half the workers in the mills and camps. The wages of all survivors, from manager to office boy, were cut 15 percent.[10] The Panic was on.

For the preceding year prices had been falling around the

world. European banks, which had underwritten much of the railroad development in America, began to call back their money. Bankers in the United States, faced with an adverse trade balance and increasing public demand for government-sponsored inflation through the purchase by the treasury of all available silver at the rate of sixteen ounces of silver for one ounce of gold, converted their paper money to gold. The treasury's efforts to persuade the financial community to accept paper currency in return for gold not only failed, the attempt increased the alarm of investors.

The secretary of the treasury admitted in April that the nation's gold reserve had fallen through the traditional support floor of $1,000,000 and made matters worse by the poorly worded pledge that the treasury would continue to buy silver notes with gold "as long as I have gold lawfully available for the purpose." The failure of the National Cordage Company on Wall Street started a bank run that amounted to hysteria, sweeping the country as stocks plummeted.

In New York money virtually disappeared. Certified checks sold at 4 percent discount. Clearinghouse bank scrip was used for exchange. Out west, things were even worse. Money was sucked east until western banks, when called on for payment by eastern creditors, had nothing to send. They begged time, begged new loans, scrounged for interest payments from their local debtors, foreclosed mortgages, only to go broke themselves. Five hundred banks closed their doors that summer. Panic petrified into a depression that

was to drag on for years. St. Paul & Tacoma asked and received permission from the Northern Pacific to delay construction of the second of the two mills it had promised to build.[11]

St. Paul & Tacoma continued to saw lumber throughout the debacle. Production slowed but did not halt, though much of the wood had to be sold at cost or stockpiled. The plant was maintained; the roof of Mill A was painted a balance-sheet red. The construction of the proposed shingle mill and planer mill was postponed, though a new trimming device invented by Fred "Dovetail" Butzer, the superintendent of the planer mill was installed. A twelve-gauge band saw, heavier than the pair of fifteen-gauge saws used since 1889, was tested successfully and eliminated problems with cutting large logs. Delays caused by hauling lumber a load at a time down the mile-long trestle from mill to ocean wharf were eliminated by dredging a deepwater channel to the stacking yard. Four new wells were sunk on the tideflats to assure the mills a steady supply of water.[12]

Griggs negotiated the purchase by St. Paul & Tacoma of a controlling interest in the Dickens Lumber Company of San Francisco with the payment to be made largely in lumber. The old company was incorporated and reorganized with a paid up capital of $50,000 and with Griggs as president. Dickens' yards in California were stocked with surplus fir from Puget Sound.[13]

New markets were pursued. General Montgomery Meigs, the U.S. engineer in charge of the government canal at

Keokuk on the Mississippi, was persuaded to order 100,000 feet of Douglas fir for a new lock gate. "Our object," he told the press, "is to give fir a test against oak. We have used large amounts of fir in barge construction owing to its reported strength, durability and the long lengths obtainable. Long timbers of white pine are not procurable in this market."

Thus St. Paul & Tacoma scraped through the bottom years of the depression. The company suffered but survived. The great problem for the owners was the collapse of other businesses in which they had invested heavily in the Tacoma area, especially the banks.

Foster and Griggs had served on boards of directors for national banks before coming west. Hewitt had been vice-president of the Manufacturers' Bank of Appleton, Wisconsin, and cashier in the Bank of Menasha. In their first interview in Tacoma in 1888 the partners spoke of their intention to found or acquire a bank. They had bought into the Traders' Bank of Tacoma, which had been organized by Henry A. Strong a month before they came to town.

Strong was a native of Rochester, New York. As a young man he became a partner in his uncle Myron's buggy whip business and later in a manufacturing firm known as Strong and Woodbury. He also invested in a company formed to manufacture the inventions of his boyhood friend George Eastman, who in 1880 patented a dry-plate process for photography and a few years later invented the first roll-film

camera, which he called the kodak. It was put on the market in 1888. Strong by that time had left Rochester for Tacoma where he established the Tacoma Foundry and Machine Company, of which he was president and treasurer. The foundry business brought him into contact with Griggs, Hewitt, and Browne. They became associated with Strong's bank before it opened its door. When Strong returned to Rochester in 1891 to become president of the Eastman Kodak Company, Griggs bought a block of Eastman stock.

Although Strong was the largest shareholder in Traders' Bank with 970 shares out of 5,000 issued, the list of shareholders in 1893 was heavy with St. Paul & Tacoma officers, their relatives, associates, and friends. Henry Hewitt had 500 shares; his father, 100; his brother Will, 100, and his friend Francis Kimberley, 100. Griggs had 100 shares and Browne, 50. Other investors included John D. Rockefeller, 100 shares; Ben Lombard, Jr., 50; and Paul Schulze, of the Northern Pacific, 10. A. N. Finch, a Rochester attorney who came west with Strong, served as president; Hercules L. Achilles, a lean, delicate accountant from Rochester, was cashier; Dr. H. C. Bostwick, the first physician in Tacoma, was vice-president; Hewitt was second-vice-president, and Griggs and Browne were among the trustees.[14]

Traders' was capitalized at $500,000. It prospered, as did most of Tacoma's twenty-one banks, during the great boom. In October 1892, Traders' reported deposits of more than $950,000. The bank moved from a small suite it occupied in the Fife Hotel building to the entire ground floor of the handsome new Berlin Building on the northeast corner of Eleventh and Pacific (the present site of the People's Store). The move coincided with a downturn in Tacoma's growth and good fortune.

As confidence waned in the city and its institutions, depositors withdrew funds from the banks. There was no special run on Traders' Bank, no incident that broke the people's trust, but the drain was steady. It increased after the Black Friday crash, increased again when the Merchants National Bank of Tacoma closed its door on June 1. In the ten weeks after Black Friday, Traders' paid out $300,000 over the counter; its deposits fell to $170,000—18 percent of what they had totalled in October.

Henry Hewitt hurried east and joined Henry Strong in a search for liquid funds. New York proved a Sahara for cash. The securities that Hewitt and Strong offered as surety for a loan had not enough appeal to tempt offers of twenty-five cents on the dollar.

On the evening of Thursday, July 20, the officers of Traders' Bank who lived in Tacoma gathered in the boardroom to review the situation. Those present included President Finch, Vice-president Bostwick, Cashier Achilles, Assistant Cashier W. G. Hellar, and Griggs, Browne, Schulze, and A. M. Stewart, another trustee. H. S. Griggs and his law partner, L. B. Lockwood, were also present as representatives of the St. Paul & Tacoma Lumber Company. The news was bad. At the July rate of withdrawal, available

money would be exhausted before the end of the month. All the trustees could do was wait for a telegram from Hewitt and Strong who had promised to wire if they struck cash. No telegram came.

At midnight the directors voted to close the bank immediately, before more losses were incurred. They hoped the closure would be temporary. A delegation went by carriage at 1 A.M. to the home of Superior Court Judge John A. Stallcup, who lived across G Street from Wright Park. They presented him with a petition asking that a receiver be appointed. The reason was embodied in a complaint signed by Lockwood, who alleged that he had $50 on deposit subject to sight draft but had been refused payment on a check drawn for that amount. His affidavit described the bank as "an insolvent institution" with liabilities that to the best of his knowledge amounted to $200,000. President Finch, acting as attorney for the bank, filed an answer admitting all allegations except insolvency. He too requested the appointment of a receiver.

At three A.M. Judge Stallcup named Addison Foster receiver. Griggs, Browne, and Percy Norton all signed $100,000 surety bonds for Foster's performance. An hour later, the delegation pounded on the door of County Clerk William Ryan's house at 917 South Jay. When he appeared in a nightshirt they presented him with the papers declaring the receivership. At 10 A.M., no telegram from New York having arrived announcing salvation, Hercules Achilles posted on the door of the bank the notice of closure:

To the Depositors and Creditors of Traders' Bank
of Tacoma:

Owing to the stringency of the money market, the continuing withdrawal of deposits, and the inability to realize immediately upon securities this bank is obliged temporarily to suspend.

TRADERS' BANK OF TACOMA
H. L. Achilles, Cashier
Tacoma, Wn., July 21, 1893

Achilles went home and took to his bed but President Finch faced the reporters who gathered at the bank. "The continuation of business is simply a question of getting ready money," he said. "Our securities are ample to pay every obligation and leave a large surplus. It has been absolutely out of the question to get ready money anywhere. . . . We should have had funds long ago had it not been for conditions which have made it impossible to secure money on any securities no matter how good. We hope to be able to resume in the near future. As I said, it is simply this question of getting ready money. If we could realize on even a very moderate amount of our securities, we would resume at once."

But they could not realize on the securities. No one was willing to loan money with the stock market still falling. The months dragged on, and Traders' remained closed. More banks in Tacoma shut their doors, most never to reopen. Of the twenty-one in town when the crash came, only four survived the year.[15]

Hewitt, Griggs, Browne, and Strong entered into an informal alliance, which pledged that every depositor and creditor of Traders' would get back every penny. As a first step they agreed that all money they had borrowed from the bank should be repaid immediately. Hewitt, who had received several large loans, paid all but $6,000 in cash and gave a new note promising the remainder on demand if it became essential. St. Paul & Tacoma repaid its loans in full. By the end of the year, $126,000 in new cash was available. Foster felt the bank could reopen if it could get a loan of $60,000 at 6 percent for a year.[16]

Syndicate members put up personally owned stocks with a face value of more than $200,000, which Browne and Griggs took to New York. With Strong's help they were able to borrow $50,000 on the pledge of securities worth four times that much at par. On January 24, 1894, the receivership was lifted and the Traders' Bank of Tacoma reopened. The men in the syndicate were betting that a resurgence in the national economy and the business that St. Paul & Tacoma could give the bank, would make operations profitable. The reopening, they felt, would help break the depression. Instead the depression deepened.[17]

The bank had funds adequate to pay all depositors, but little working capital. They were able to reduce the amount owed depositors from the $170,000 outstanding when the bank shut its doors to $55,000, of which $18,000 was on deposit for the City of Tacoma and $5,000 for the Northern Pacific. This reduction depleted the bank's cash reserve.

Some of the firms that had done business with other banks that failed were interested in working through Traders' but the officers could not encourage them since they did not have enough money to serve their old customers properly. It was a trying business.

President Finch resigned in April and returned to the private practice of law. His position was not filled. Vice-president Henry Strong was in the East. The cashier ran the bank with advice from the trustees.

On Friday, May 18, Hewitt and Browne reviewed the situation. The books were in order but it was costing $40,000 a year in rent, insurance, taxes, clerk hire, and miscellaneous expenses to keep the bank open. There was neither profit nor the prospect of profit in the near future. "There is no use running a business for the mere purpose of making ends meet," said Hewitt. They decided it was time to get out of the banking business.

Saturday morning Hewitt and Browne appeared before Judge Stallcup again. They filed a complaint that the bank owed a large amount of money and that while assets were more than sufficient to pay every claim in full eventually, it was impossible to realize upon the assets at the moment. They asked the appointment of a permanent receiver to close out the business. The judged appointed Leonard Howarth, Hewitt's private secretary, as receiver and directed him to proceed "with all expediency" to realize on the assets, to submit a report every two months, and whenever he had at least $10,000 at hand to declare a dividend and never to accumulate a sum equal to $20,000 without declaring a dividend.[18]

A native of England, Howarth had served an apprenticeship there in the printing trade before coming to America with his brother William. They settled in Wisconsin, where Leonard became associated with the Hewitt interests. He came to Tacoma in 1891. A quiet, clerkish man, partially crippled by infantile paralysis, notoriously shy (he never married), Howarth attracted little attention except among close associates. They recognized in him a man with exceptional fondness for detail work and with a shrewd, unsentimental sense of financial opportunity. During Addison Foster's period of receivership he had left the actual management of Traders' affairs to Howarth. Now as receiver in his own right, Howarth carried out the court's orders but gave the syndicate as much time as possible to find money to repay the bank's debts without selling the sureties on a depressed market.

The second closing of Traders' jeopardized the financial position of the St. Paul & Tacoma officers. It was not just that Griggs, Hewitt, and Browne had put up security for the last loan to the bank and stood to lose stock of great potential value if foreclosed. More than that was at stake. They could lose the confidence of eastern financiers in their judgment and integrity. This would not only make it difficult for them and their company to borrow money in the future, it could retard recovery throughout the West.

Griggs outlined the problem in a letter to A. Trow-

Leonard Howarth came to Tacoma to work for the Hewitts but became a major figure in the financial operations of the lumber company (from American Lumberman, *May 21, 1921)*

bridge, cashier of the National Bank of North America in New York. "So long as the moneyed men of our country have not full confidence that the public and the business community intend to pay one hundred cents on the dollar, there is no doubt in my mind that we will be unable to obtain any eastern funds in our western country for investment. I can assure you that the rest of the syndicate feel it as pointedly as I do and are willing to do anything in their power to meet this indebtedness."[19]

Griggs undertook the task of seeing that repayments reached New York banks in time to prevent the sale at depression prices of a total of $700,000 in securities against debts of $200,000, and of seeing that every depositor got his money back. It took three years.

Guiding St. Paul & Tacoma through the greatest depression the nation had known and cleaning up the mess left by the collapse of Traders' took all of Griggs's energy. He liquidated most of his outside business interests except for the Beaver Dam Lumber Company in Cumberland and real estate holdings in the Midwest, which it would have been foolish if not impossible to unload on a depressed market. As it was he accepted losses of more than a quarter million dollars. In addition he took $10,000 cash as replacement of a loan he had made Henry Strong rather than accepting an offer of stock in Eastman Kodak.[20] It amused him later to point out that within ten years the shares he had rejected out of need for cash were worth half a million dollars.

Reducing expenses and staying within income became an

obsession. Griggs's every letter to his young son, Theodore, who was attending Yale, preached the need of economy. ("It seems the amount you require runs considerably over $100 every month and with times as close as they are now, we will have to begin cutting down expenses. Unless you can make yours less, it may be impossible to return you to college next year. Undoubtedly you are not spending any more than any of the other boys, but the business interests of the country require immediate reform in this matter and all expenses will have to be cut off outside of what is necessary for actual living, clothes and tuition. Please act on this immediately. Acknowledge receipt of this. Affectionately yours . . ."[21]

He wrote the *Tacoma Ledger* to dispute a seventy-five cent bill, wrote the gas company ordering them to take out their line unless they could reduce charges, wrote to his tailor in St. Paul ordering a new suit and coat "equal to what I would get from Brooks Brothers in New York but much less expensive." He backed out of a twenty-five dollar pledge to the Church of the Holy Communion on the ground that the pastor they had hired was not, as promised, "a man of great influence, an orator, second to none in his profession in the east."[22] He refused contributions to Democratic candidates. "I feel sorry that I cannot comply with your request," he wrote to a Tacoma friend who sought county office, but

. . . circumstances which I have no need to go into are such that there is no corn in the crib, and in fact I have not what I need to meet personal expenses. Until there is a change in monetary conditions I will not be able to assist even those I count on as my very best friends, in a financial way. Not only the banks but those that I have been guarantor for have compelled me to use all of my surplus cash the last eighteen months to meet their obligations and I have had to borrow and obtain money where I could to the best advantage even to do that. Property at present time makes no difference in relation to raising money, and the circumstances I have recounted will show you why it is impossible to comply with your request.[23]

He declined to enter a St. Paul business venture with his son Chauncey Milton, though he felt it certain to prosper. He unabashedly ended a business letter to James Hill with a request for free passes on the Great Northern for his college-age children's transportation.[24]

While counting pennies, Griggs concentrated on revitalizing three bankrupt businesses in which he had bought stock and which owed significant amounts of money to Traders' Bank. They were the Puget Sound Dry Dock Company, the Pacific Meat Company, and the Crescent Creamery. If restored to financial health they could repay their loans to the bank. Failing that, they could be foreclosed on, sold, and the money used in paying off Traders' obligations.

Crescent Creamery, in which Griggs, Foster, and Hewitt had bought stock because the organizers were acquaintances from St. Paul, proved beyond redemption. Griggs decided that it had been "conceived in fraud." He wrote sorrowfully to his eldest son that "I shall never again become financially involved with anyone I have not known well for at least half my life." Though the Creamery never reopened, considerable money was realized for Traders' through the sale of its handsome building on Ocean Wharf. It became the Pacific Cold Storage Company.[25]

The Pacific Meat Company had been started in the 1880s by Tacoma, Portland, and Seattle interests. Its packing plant was at Meaker Junction, up the valley a mile east of Puyallup. The site had the advantage of being convenient to a good area for penning animals but suffered from the high rates for short-haul service charged by the Northern Pacific. While Pacific Meat was in the hands of receivers, some of its buildings burned. Early in 1895 Griggs learned that the receivers had decided to rebuild in Seattle. Griggs pleaded the cause of Tacoma. He was told a final decision was to be made in two weeks. If he could put together a solid proposal in that time it would be considered.

Griggs moved swiftly. He persuaded the Tacoma Land Company to give two and a half acres on the boot adjoining the St. Paul & Tacoma boarding house to Pacific Meat because "the number of employees at the plant will range between 65 and 120, which means an addition to our population of at least 500 people, all of which, if located in this city, will be compelled to purchase lots for the homes of their families." Besides that, the packers' shipments provided the Northern Pacific with $100,000 a year freight revenue. Since the Northern Pacific was the parent company of the Tacoma Land Company and Tacoma was losing population at the rate of 200 persons a month, these proved persuasive arguments. Isaac Anderson, the land company manager, agreed that the land would be available if Pacific Meat built in Tacoma.[26]

Next Griggs won approval from the Northern Pacific for a revised freight schedule that would save Pacific Meat $10,000 to $15,000 a year. The railroad agreed to contribute $5,000 to provide a spur to the plant site. St. Paul & Tacoma offered to sell at cost the lumber needed for the five-story, 150-foot-square packing plant.

Ten days after Griggs began his campaign, Pacific Meat not only accepted his plan for a move to Tacoma, they made the Colonel president of the company. He loved the food business, as his earlier success with Cooper and Griggs in St. Paul indicated. He clearly enjoyed building up the sales by influencing other lumbermen to buy food for their logging camps from Pacific Meat. He created a new line of ham, bacon, and canned meat to be sold under the brand name of "Imperial," and a fertilizer called "Webfoot," and found markets for them in Russia, China, and the Hawaiian Islands.

The first year's profits were $25,000. Griggs believed the company could make $100,000 a year if there were enough capital to assure maximum purchases of livestock. Nevertheless he kept trying to sell the plant to one of the major midwest packers (Armour, Swift, or Cudahy), since nearly half the stock in Pacific Meat was held by Traders' and a sale would provide enough funds to pay off most of the

Crescent Creamery Company, Griggs decided, was "conceived in fraud" and beyond saving. The building was turned into a cold-storage operation (from Spike's Illustrated Description of the City of Tacoma, *1891; Tacoma Public Library)*

bank's debt. His correspondence with the midwest packers contended that Pacific Meat "has an advantage in the pork packing business of nearly two cents a pound over eastern houses, considering the prices they buy their hogs at, delivered here, and the freight from there to our coast on prepared meats." No sale. The major companies nibbled but did not bite. Griggs stayed on as president of the company, presiding over an operation that returned steady profits and helped clear the Traders' debt. In 1903 the plant was purchased by the Carstens brothers of Tacoma.[27]

The Puget Sound Dry Dock posed different problems. Captain DeLion, who had supervised its construction at Port Hadlock and managed its operation after it was towed to Dockton on Maury Island, cracked under the strains imposed by the depression. After brief hospitalization in Seattle, he took his own life in 1894. Billy Fife, the largest stockholder in the company, found himself, like so many other Tacoma area capitalists, land rich at a time when land could not be sold. He stood by helplessly as the company slid into bankruptcy because shipping on Puget Sound decreased more than 50 percent and the captains of such vessels as did call could not pay for work that was done.

The courts appointed Leonard Howarth receiver for the dry dock company as well as the bank. Griggs tried to arrange a sale. The actual cost of building the dock and its supporting equipment at Quartermaster Harbor was $450,000. Independent experts appraised it in 1895 as worth $325,000. Griggs proposed that the syndicate members, who controlled a majority of the stock, foreclose, then sell the facility to San Francisco shipping interests for $175,000, which would then be applied to the Traders' obligations. The deal fell through, so Griggs ran the dry dock himself. It broke even the first year, then made a profit each year until the early 1900s when it was sold to the Manson brothers who operated a pile driving company. They had the dry dock towed to Seattle in 1909.[28]

While Griggs was doing double duty for the syndicate and St. Paul & Tacoma, his work was complicated by the bankruptcy of the Northern Pacific and the suicide of Paul Schulze, the railroad's land agent, whose romantic pursuits had led him into six-figure embezzlement. Griggs had not admired Schulze. With uncharacteristic coldness he described the suicide as "the end which the better class of citizens generally think such men are liable to." He went on to say that Schulze "was not worthy to be classed in the element of men of reliability in any sense of the word . . . I think we are better off without him or any of that class of men in our midst."[29]

Still Schulze's death removed the official with whom Griggs had most often dealt in interpreting the wording of the St. Paul & Tacoma contract with the Northern Pacific. Receivers appointed to administer the railroad's affairs sought to reinterpret every clause in a way that would wring the most money from the lumber company. Griggs managed to demonstrate the inequity of most of their proposals. Finally he arranged a face-to-face meeting with the new president of the railroad, a representative of Pierpont, Morgan, & Company, and the chairman of the reorganization committee "and after three hours in consultation and discussion on our matters, we came to an amicable and satisfactory agreement, and have entered into a supplementary contract agreement covering some few points which occasioned a little difficulty between the road and ourselves." Among other things, the Northern Pacific agreed not to insist on construction of the promised second mill on the tideflats until times were better. It was a considerable victory, but the drain on Griggs' time and energy had been considerable.[30]

Politics provided another distraction. In 1896 the Democrats split over the "free silver" question. Griggs's hero, Cleveland, who favored remaining on the gold standard, lost control of the party. William Jennings Bryan stampeded the convention in Chicago with his cry that "You shall not crucify mankind upon a cross of gold." Bryan was nominated for the presidency and the platform called for the purchase by the treasury of all silver offered to it with payment to be at the rate of one ounce of gold for sixteen ounces of silver.

In letters to his children before the Panic, Griggs had indicated an open mind about the merit of stimulating the economy by such a method, but his problems in the depression ended that flirtation with unorthodoxy. He considered the Chicago convention "a fraud"; he warned his son Milton that "it would be hazardous for us to spread out in any way until after the election . . . it is better to be conservative on everything than to take any chances which may bring peril in the end to any concern which is doing a large business."[31]

When the Republicans nominated William McKinley and came out for the gold standard and a protective tariff, he bolted the Democrats and supported the Republican presidential ticket in every way he could. After the election he congratulated McKinley's manager, Mark Hanna,[32] for "wonderful generalship" and apologized that "we were not able to carry this state for McKinley and Hobart." He warned Hanna that "we have more irresponsible men in both parties of this state than in any other in which I have ever lived, and it will require better judgment to make selections which will assist in building up the party for Sound Money and Protection and the general good of our people than in any other section of the country."[33]

In 1897 the courts decided that out-of-state stockholders in the Traders' Bank were personally liable for claims from creditor banks. Griggs took the lead in forming a Settlement Company, made up of the thirty persons most at risk. He served as its president with Strong as vice-president, Browne, secretary, and Howarth, treasurer. The debt had by this time been greatly reduced. The Settlement Company in 1898 was able to pay off the last of its obligations and reclaim the stocks that had been posted as surety.[34]

On October 22, 1898, Griggs wrote a triumphant and moving letter to his friend and financial guru, Henry Dimmock, about the business, "which was entered into about

ten years ago, when we all felt so confident of what would be the result of our investment in this far Western country."

We have passed through the hard times and are on the upward trend, for it has been a herculean task to overcome obstacles which have been thrust in our path since we came to this beautiful section.

. . . we started out by paying in assessments . . . the purchase of a large tract of land from the Northern Pacific, upon which due care was taken in making our contract with the railroad company, at your suggestion employing the best counsel that could be had in your city, which proved to be a very wise thing for our Company, as the Northern Pacific, of course, finally went into the hands of receivers, as you at one time thought might so happen when discussing the matter with President Oakes, and he so earnestly protested against any probable occurrence that could occur to the Northern Pacific of that kind. By close watching and looking after that interest, we have finally overcome that difficulty and got ourselves into splendid condition.

At the time we made the investment, common lumber was selling in this locality at $11 per M. From the first year it commenced to drop and went to as low as $6 for the same grade in 1893. You can readily see the effect it had upon a large enterprise of this kind which would mean a difference of $1,000 a day to the income of our property, and with ninety per cent of our men in this country at that time broke.

The writer had assumed and guaranteed individually over $500,000 to assist enterprises in going forward and saving them from bankruptcy, and was then on the bonds of the Northern Pacific to the amount of $200,000. You will remember it was a pretty blue time when last I visited you in New York, relative to the Traders' Bank, when I fixed up the deal that was to protect your own bank and other banks that we had guaranteed and deposited the securities with our friend Mr. Trowbridge, then Cashier of the bank.

I felt I had a great load upon my shoulders at the time, and the comforting words you gave me, together with your directors, gave me good courage to carry out the very idea which has always been in my mind—to meet every liability, whether pledged for others or in connection with myself in doing business. This matter has been fully carried out and every one of the obligations wiped away, and I feel like a new man, emerging from a terrible fire which had scorched me somewhat but left me in good shape for the future work I had to perform.[35]

New Markets

THE ST. PAUL & TACOMA operation had been planned for shipment by rail, but with freight rates remaining a problem the seaborne trade with California and the Pacific rim countries remained vital to its success throughout the 1890s. Henry Hewitt, following his Everett adventure, assumed responsibility for developing this market.

In 1896, Hewitt, accompanied by Charles Nelson, a California ship's captain and lumber dealer, set off aboard the sailing vessel *Celtic Race* for a visit to Australia and the Far East. The *West Coast Lumberman* predicted that by the time Hewitt returned "he will be able no doubt to speak several languages, not including christian, profane and the language he learned when operating in saw mills and timber in Wisconsin."[1] Griggs expressed the hope that he would be able to sell some lumber for the Broken Hill mining development in Australia, ties for railroads in China and Japan, and establish ties with strong companies to distribute St. Paul & Tacoma lumber.[2]

Hewitt's style as an ambassador of trade was unorthodox but effective and his visit to Sydney attracted international attention. Colonel Bell, the U.S. consul in that city, gave a banquet for Hewitt and Nelson, which was attended by the premier, the attorney general, and several members of Parliament. Called upon to respond to the toast, "Our American Brothers," Henry delivered what one listener described as the most humorous speech given in Sydney since the visit of Mark Twain.

I had no idea of finding anything like this, . . . a great city with so much money in fine buildings and such polite society. I did not expect to see people arrayed in claw-hammer coats at an evening supper. Why I never wore a spike-tailed coat before in my life (laughter)—and this is not mine (roars of laughter).

I came over here to see what you were doing, how much lumber you would buy, and I found a great country. Over there we get up at four o'clock in the morning to hunt business, and when we get a surplus we go abroad to find a market. Here I am.

I was not born American, I was born in England. I could not help it. (Loud laughter) If I had been born six months later I could have been born in America. My father was a contractor, so in early life I contracted the habit of attending to business. I helped to develop Wisconsin by removing the trees, cutting lumber. Then I helped to develop Michigan until her trees were pretty scarce. Then I came west to grow up with the country, and I found the best trees in the world, and I am developing Washington. I could not get further west without taking to water, so I came over to see how much lumber you wanted, and I like it so well, if I was not so darned old, I would come and help develop Australia. (Cheers and laughter)

After noting that Australia had "everything but pine trees," which he offered to supply them, Hewitt turned to economic theory. "You have everything here: good country, good people, good climate, good harbor. But climate won't yield 2 percent. You have coal, iron, wood, copper, lead, soil and everything. Why don't you make your own goods? You make more mistakes than you do pig iron."

Whereupon Hewitt launched into a speech on the merits of a protective tariff to the intense amusement of the premier, who was Australia's leading advocate of free trade. Hewitt said he would protect everything, shut out or tax everybody who wanted to come in—except lumber people— and build up an Australian empire.

Referring to the policy of Australia in borrowing money to develop its resources (and perhaps recalling his activities in booming Everett), Hewitt said, "That's right, borrow all you can get. Borrow and build railroads, build factories, build shops and make your own goods. If you can't pay the interest the lenders can't take the improvements, and if they do they will have to come and run them and then you will have the improvements and the skilled people too. (Laughter)"

At this point Hewitt made a broad gesture and upset an uncorked wine bottle. He studied the table for a long moment. "Well, I've enough in me," he said, setting the bottle back upright, "but it is the surplus we should look after. We must save the surplus. It's the surplus we trade in, and it's my surplus lumber I want to sell to you people. That's what I came over here for."

Hewitt's mixture of blarney, brass, and candor upset Captain Nelson, who wrote a complaining letter to Colonel Griggs, but St. Paul & Tacoma received orders from the Broken Hill Mining people for several shiploads of no. 2 lumber, the grade then in greatest surplus at the mill.[3]

From Sydney, the travelers went to Melbourne and Adelaide, then to Canton where Hewitt made a large sale, but at prices Griggs felt were regrettably low. He also found, however, a specialty market for spruce to be used in the manufacture of tea chests. In Yokohama he made a connection with a local distributor he predicted "would take all of our yard and common lumber that we do not dispose of otherwise," and sold railroad ties for a line being laid across Hokkaido. He seems to have gone on to Korea and Russia, though no sales were reported from those areas.

In August of 1897, Griggs wrote a friend to say, "Our Mr. Hewitt has just returned from his extended trip to Australia, China, Japan and even to Russia . . . very much elated. We are loading one vessel for China now and another will commence next week. We have six vessels at our dock and are sending out from six to fifteen cars of lumber

Full-rigged ships used for hauling lumber were difficult to load. Stern loading was one improvisation, but it was time-consuming (Historical Photo. Coll., UW Library)

and other materials daily from the yard, so you see we are on a little different basis than when you were here."[4]

New markets were developing for lumber all around the Pacific rim. Chile's victory over Peru in the War of the Pacific (1879–83) opened the way for the exploitation of the copper and subsidiary mineral resources of the Antofagasta and Tarapaca provinces. That meant more railroads, more ties, more bridges across the gullies of the Andes, nearly all built with lumber from the Cascades and Olympics. The discovery of silver, lead, and zinc at Broken Hill in Australia in 1884 had spurred the industrial boom Down Under. Railroad construction in North China demanded more lumber than Asia's mills could supply. Connection of the Santa Fe Railroad with the Southern Pacific in 1882 touched off the never-ending migration of Americans to Southern California. Douglas fir dominated the market in all these areas.[5]

New ships were developed to carry the lumber. In the early days of Puget Sound sawmilling, the product was carried in vessels left over from the California gold rush. These craft had been built for general cargoes and, though ill-suited for the special needs of the lumber trade, they were available, they were cheap, and so they were used.

In 1875 the Board of Marine Underwriters in San Francisco, worried about the obsolescence of west coast shipping, agreed to insure ships built of Douglas fir if they met rigid specifications. Local yards began turning out vessels specifically for the lumber trade.

First came the coastal lumber schooner, two- or three-masted. Schooners could sail closer to the wind than square-riggers, and maneuver more easily, which was important in entering many of the small bays and coves where northwest mills were located. Because of their light overhead rigging they required fewer crewmen, which meant a smaller payroll. The coastal lumber schooner evolved into a flat, single-decked vessel, broad of beam, long in the bow, square in the stern, with oversized hatches that permitted easy stowing of long timbers. Since it had no second deck to add rigidity, the schooner was built rather flat, which decreased cargo space, but owners compensated by piling the deck so high that schooners sometimes put out with no freeboard at all. Old-timers yarned about going to sea with the deck four feet under.

For long hauls across the Pacific, shipyards turned out the great schooners, vessels of four, five, or even six masts. These huge craft required two decks for structural safety. This complicated loading, but the great schooners needed fewer crewmen per ton of cargo hauled than any other deep-sea vessel. More than fifty four-masters were built in the Northwest in the last two decades of the nineteenth century. The first five-master was launched at Cook Bay in 1888, and by 1900 they were making six-masted vessels.

The great schooners had a weakness. They were not as

The coastal lumber schooner was developed to facilitate loading and cut labor costs (from author's collection)

fast as square-riggers in a following wind, a deficiency important when riding the trades on long hauls across the Pacific. Some designers added yards on the foremast so that a square sail could be set to windward to pick up following winds. Others rigged their vessels as barkentines, square-sail forward, schooner rig on the other masts. This required more manpower, but the barkentine was faster and considered by some the best vessel for runs to China and Australia.[6]

And there was the vessel which came to be known as the *Everett G. Griggs,* a most unusual craft. Built in Belfast in 1883, iron-hulled and rigged as a four-masted bark, she sailed first as the *Lord Wolseley* and tramped around the world until sold to German interests; they renamed her *Columbia* and assigned her to the lumber trade in the Pacific. In 1902 she lost three masts in a storm off the Pacific and limped into Esquimault on Vancouver Island in such bad shape her owners sold her as a hulk. For more than two years she worked as a barge, lightering ballast. In 1904 she was purchased by Chauncey Griggs and some associates, towed to the Moran yards in Seattle, and refitted. When she was warped out of the rigging dock early in 1905 after $65,000 worth of repairs, she was the *Everett G. Griggs,* the world's first six-masted barkentine. (Only one other ever sailed.)

Registered Canadian to escape the more stringent U.S. shipping regulations, the *Everett G. Griggs* was put into the Australian trade. Her hold could take 100,000 board feet—two days' cut when St. Paul & Tacoma was under full steam, a statistic that pointed up the advantage of quick passages even for cargoes as non-perishable as lumber. She left Tacoma in August, 1905, under Captain G. E. Delano and returned in five months, hull-down with Australian coal. In the next few years she became a familiar sight on the sound, but the economics of the carrying trade did not favor sail over steam after 1900. It proved no advantage for a mill owner to run a ship. The *Everett G. Griggs* had to scrape for cargoes. In 1908 she carried barley to Europe from San Francisco and returned with less than a full cargo of general merchandise. The following year she showed up in Puget Sound with some pig iron for the Alaska Junk Company, then lay idle on the waterfront for months. She was sold in 1910 for about what it had cost to buy and refit her. The new owner, Captain E. R. Sterling, named her the *E. R. Sterling.* She is said to have earned a million dollars for him on high-risk voyages during the First World War. She tramped the seas until 1927 when she was damaged rounding Cape Horn with Australian wheat for England, then dismasted in a hurricane off the Virgin Islands. Towed to England she was judged useless and junked at Sunderland, about 250 miles from the port where she had been launched forty-three years earlier.[7]

Meanwhile, a new market had unexpectedly developed. On July 15, Puget Sound newspapers reported the arrival in San Francisco from the Bering Sea of the *Excelsior* with word

The iron-hulled Everett G. Griggs *was the world's first six-masted barkentine. She could carry 100,000 board feet of lumber, the equivalent of two full days' cutting* (Tacoma News Tribune)

of a major gold strike in the Yukon Territory of Canada. Two days later the steamer *Portland* docked in Seattle to discharge her legendary "ton of gold" and a contingent of sudden millionaires. The rush was on to the Klondike.

Going north did not prove rewarding to the St. Paul & Tacoma people who tried it. Henry Hewitt visited the Klondike briefly and decided that while gold was certainly a commodity everyone wanted, it was too late for him to get in early. Herbert Griggs, the attorney, formed a partnership with the Tacoma city treasurer, William S. Sternberg. They chartered the new and elegant *Charles Nelson* from Hewitt's recent traveling companion at $500 a day and sailed aboard her for St. Michael at the mouth of the Yukon, with a load of priority freight for the U.S. Geodetic Survey and a full house of passengers willing to pay extra for speed. The *Charles Nelson,* however, proved to be as slow as she was handsome. All bonus payments promised for quick delivery were lost. Griggs could find little cargo for the voyage home. He returned to lawyering, richer only in experience.

C. H. Jones was sent north by the company to establish lumber yards in Dyea and Skagway, the boom towns at the head of Lynn Canal. Everyone wanting to go over the Chilkoot or White Pass to the interior had to funnel through Dyea and Skagway, so profits seemed certain to anyone on hand early with building materials. St. Paul & Tacoma's first shipment failed to make it ashore. The bark *Canada,* at anchor in Dyea harbor, was blown loose in a January gale. Men who found her adrift forty miles away successfully claimed her and her cargo for salvage. The second ship was delayed at the docks in Tacoma—there was a shortage of longshoremen created by the departure of so many able-bodied adventurers for the goldfields. By the time the lumber reached Skagway the market was glutted. The St. Paul & Tacoma agent aboard did not bother to unload. He took the lumber back south to Wrangell and sold it at a slight profit.

The loss of the *Canada* was a special problem for one Tacoman who had come north on her, Hattie M. Lockwood. She had been the wife of Herbert Griggs's first law partner, Charles Lockwood, but he had abandoned her, their infant son, his law career, and Tacoma to disappear with a young woman who had come to Tacoma to give French lessons.

Left on her own, Mrs. Lockwood hit on the idea of going north and establishing a convalescent home for exhausted prospectors. She found another woman interested in the concept of a "hospital without doctors." They assembled supplies and sailed for Skagway, living aboard the *Canada* while it waited a chance to unload. When the storm drove the ship down the canal they were rescued but all of their rest-home equipment was lost. Her companion went home, but Hattie opened a boarding house in Skagway and kept it going through the great years of the rush. (On her return to Tacoma, Hattie operated a boarding house for loggers; her establishment was reputed to be the most genteel of its kind: lace curtains, flowers on the table, and clean sheets every week. Later she went to Hawaii, married a wealthy man, and became known for her contributions to educational institutions.)[8]

Though the Klondike rush failed to bring sudden fortune to any of the St. Paul & Tacoma people who went to Alaska and the Yukon, the company benefited from sales to local shipyards and construction companies, and from the upturn in the regional economy promoted by the outflow of materials and the influx of gold.

The Spanish–American War added to the boom. Colonel Griggs, who had not forgotten his own Civil War experiences, was aware that a war would spur industrial activity, but he watched with dismay the newspaper campaign for military action in Cuba.

"There is no doubt if the jingoes and irresponsible men of our country can plunge us into war, they will do so," he wrote in a letter to Hewitt. "What in the world does the United States want of any more territory, unless it is for coaling purposes in case we get into war with foreign countries? War means death and destruction, and what it may lead to no one can tell. When our troubles commenced with the South in the late Rebellion, it was thought by many men in the North that they could be wiped out in thirty days. Little these men knew what war meant and what it would take to bring about order and peace to our country."[9]

Nor, when the war did come, was Griggs swept away by enthusiasm. "How much good the war will do us is more than anyone can tell," he observed in a letter to his daughter Anna, "but undoubtedly it will in the end create a very extensive navy and require us to have a large standing army, especially if we acquire the territory we take, including the Philippine Islands, Cuba, etc."[10]

His son Theodore wired from St. Paul to ask, "Have you any objections to my enlisting in the army?" The Colonel responded with an old soldier's advice: Don't step forward. "You have started out to become a groceryman, and until your services are very much needed by the country you had better defer tendering them."

Theodore wired again to say that if he answered the call, he would go as a commissioned officer. His father told him to get it in writing "or else you may be left out in the cold and be only a private. The opportunity for rapid advancement in the army undoubtedly will be slow, unless there is a large loss of life by disease or otherwise, which is as apt to take you as anyone else. Consider this well before you go in. Should you conclude finally to enlist, you will have my best wishes, and I should do anything that was possible for you, provided you returned."[11]

The Colonel was proud but sad when Ted volunteered— and unabashedly happy when the war ended before either Theodore or Everett, who was an officer in the National Guard cavalry, could be exposed to enemy fire or, what their father feared more, yellow fever.

After the war was won, Griggs worried about the effect of empire on America. "The only thing that in my judgment will be not so beneficial to our country," he wrote

Anna, "will be the extensive domain that probably we may retain on account of our having captured certain portions of their territory." To his son Chauncey Milton he said, "If our country undertakes to hold these conquered lands, they will require a standing army equal to the one we have in the field. I certainly hope that a staid Government, which will be satisfactory to not only the people in these distant lands but to our country and to all other foreign nations, will be formed, and relieve us from the responsibility."[12]

The national debate over the status of territories liberated from Spanish rule found Griggs in rare disagreement with President McKinley and with his dearest friend, Addison Foster, who had just been elected to the United States Senate. The controversy was climaxed by a decision reached by the president in prayer and solitude.

"I walked the floor of the White House night after night until midnight," McKinley told a delegation of Methodist missionaries who favored annexation, "and I am not ashamed to tell you, gentlemen, that I went down on my knees and prayed Almighty God for light and guidance more than one night. And one night late it came to me this way—I don't know how it was but it came . . . that there was nothing left for us to do but to take them all, and to educate the Filipinos, and uplift them and civilize them and Christianize them, and by God's grace do the very best we could by them, as our fellow-men for whom Christ also died. And then I went to bed, and went to sleep, and slept soundly, and the next morning I sent for the chief engineer of the War Department, our map-maker, and I told him to put the Philippines on the Map of the United States, and there they are, and there they will stay while I am President."

Many Philippine nationalists objected to such uplift. Emilio Aguinaldo proclaimed a Philippine Republic, and as Griggs had feared an American military campaign was mounted to establish possession. The war against the Philippine nationalists cost far more money and took more lives than had the Spanish–American War, but it was not without some benefit to St. Paul & Tacoma and the Puget Sound economy. The June 1900 issue of *West Coast Lumberman* noted, "The St. Paul and Tacoma Lumber Company made quick work loading the big cargo of lumber for the government forces at Manila." It was the first of many loads.[13]

A Senator from St. Paul and from Tacoma

IN JULY of 1898 the courtly Kentuckian Colonel Jim Knox of Puyallup, who was a minor figure in the Republican party of Pierce County, suggested to Addison Foster that he seek election to the United States Senate.

This seems to have been a new idea to Foster, who was no self-starter in politics, but one that interested him, especially as Knox reviewed the political background in Washington State. The legislature was dominated by Fusionists, as the alliance of Democrats and Populists carried into Olympia by the 1896 flood of discontent with the economy called themselves, but the Fusionists were drifting apart now, carried in different directions by the forces of economic recovery generated by the Klondike gold rush and the start of the Spanish–American War. Eighteen ninety-eight looked like a winning year for the Republicans. If they controlled the legislature, a Republican would almost certainly be elected to the Senate.

Both Washington senators were Republican, but both were from eastern Washington. The seat up for contest was held by John L. ("Little Johnny") Wilson of Spokane, a diminutive chap with a large mustache, a small bankroll, but an understanding with James J. Hill that assured him campaign funds.

Wilson's eastern Washington base was bad enough for Pierce County (Tacoma papers referred to him as "the coyote statesman"), but friendship with Hill was the unforgivable sin. Hill meant the Great Northern and Seattle. What was needed, Knox assured Foster, was a Republican from the wet side of the mountains, someone not beholden to Hill, someone who would concentrate on getting harbor improvements for Commencement Bay.[1]

Foster canvassed such local Republican powers as Stanton Warburton, Jim Wickersham, Charles Murray, Mayor Nickeus, George Browne, and City Council President Percy Norton, as well as the pinochle-and-domino activists at the Union Club. They gave encouragement, though some of the professional vote seekers among them pointed out certain liabilities in a Foster candidacy: he had taken no active part in Washington politics, and he lived half of each year in St. Paul or Cumberland, Wisconsin.

The decisive influence was probably Chauncey Griggs. He believed it the duty of stable businessmen to play an active role in politics, not merely by financial contributions but by serving in office. He had encouraged Browne's candidacy for the state legislature, Norton's for the city council, and twice had actively sought a senate seat himself as a Democrat. Though Foster was a Republican their political differences were few. Over their evening pinochle they agreed that it would be good for the country if Foster were a senator, and not bad for St. Paul & Tacoma.

On August 5, Foster confided the first mention of his possible candidacy to his daily journal: "Talk about town for me to run for U.S. Senate." The next day: "U.S. Senator question being strongly talked of." And the next: "I was interviewed by Ledger as to my candidacy for U.S. Senator." On August 11: "This A.M. the Ledger and Post Intelligencer say I am a candidate for U.S. Senator. Left for Yakima."[2]

The headline on the *Ledger* account read: "FOSTER IN THE RUNNING . . . STRONG FIGHT TO BE MADE. BELIEVED TO BE THE MAN AROUND WHOM LOCAL ANTI-WILSON MEN WILL RALLY." The quote from Foster was hardly a trumpet call. "If I am convinced that there is a chance of being elected, I shall certainly be a candidate. A man does not often refuse such a position. I am not inclined to throw a senatorship over my shoulder."[3]

The *Tacoma News,* at the moment in flux between ownerships, was less kind. "REPUBLICANS ARE MYSTIFIED," said its headline. The story was couched in terms of highfalutin bemusement:

A new Warwick has sprung into the political arena, drawn his bow across the horse fiddle of candidacy, and challenged the host of Republican aspirants to wage him battle for the senatorial toga. He is Addison G. Foster, vice president of the St. Paul & Tacoma Mill Company and hitherto a gentleman of good repute. The dulcet tones of the Lorelei of Politics have struck the tympanum of his ear with pleasing vibrations until his charmed fancy has magnified them into the well of vox populi—in short, the politicians who have induced him to "come out" have made him believe that the voice of the people is demanding him with thunderous accord. He hesitated at first, then "saying he would never consent—consented." Just why Mr. Foster should be trotted out when so much available timber was lying around loose seems very mysterious indeed.[4]

The general reaction did not discourage Foster. "Tacoma comes out for me for U.S. Senator," was his summary in his journal the day after the news of his candidacy was published. A few days later he reported that "about 1,000 men came to me to tell me they want me for U.S. Senator." He had indeed listened to the Lorelei and was captive.

With Foster committed to the campaign, the professionals took over the work of lining up support. Jim Knox moved through western Washington courting potential Republican legislators and those who might influence them. Percy Norton, who in lumber affairs had proved to be a quiet and able negotiator and on the city council an effective campaigner, worked the Seattle area with great delicacy.

Since a Tacoma candidate was not likely to have much

Senator Addison G. Foster, the first senator from Tacoma (and St. Paul) (courtesy of Mrs. William Pettit)

strength up north, Norton's strategy was to make Foster acceptable as a compromise if Seattle's favorites faltered. He avoided direct attack on Wilson, whose services for Hill gave him a Seattle base, and he let the Wilson people carry the fight against former Seattle mayor Tom Hume, who also sought the senatorship. Norton did his quiet best to get Hume and Wilson at each other's throats.

Foster's role was to travel about the state being amiable and old-shoe, calming doubts, kissing babies, shaking hands. It worked. "He is a man who would much prefer to eat a dish of steaming baked beans in a logging camp, surrounded by jovial, story-telling loggers, than to wear a dress suit at a Belshazzer banquet and listen to formal toasts," wrote one interviewer. "He would not fit a dress suit. The everyday business suit he wears fits him but little better than a paper bag fits a dime's worth of bananas. There are few men who can entirely subordinate the clothes they wear. Mr. Foster is one of them. You might talk with him for half an hour and if you retained any impressions at all of how he was dressed it would most likely be a dim consciousness that his trousers bag slightly at the knees and that his cuffs are big enough to reveal underneath unusually heavy underwear, turned back at the wrist."[5]

Addison's son Harry, a Yale graduate, often substituted at more formal meetings and delivered, whenever occasion could be found, the family response to the proposed toast, "Water—the Purest Creation of Providence":

"My friends, I have seen water glisten in tiny teardrops on the sleeping lids of infancy. I have seen it trickle down the dimples of youth when soft lips yield to love's caress, and on the whitened cheeks of age. I have seen it drip like a shower of gems from the blades of grass on the resplendent dawn of a new day. I have seen it tumble down mountain sides in cascades as fleecy as a bridal veil. I have seen the seven seas on whose bosoms float fleets of all nations, and the commerce of the world. I could dilate more in extenso. But as time waits for no man and pressing importunities of the present occasion preterminate, my friends, I must hasten to say to you that water as a beverage isn't worth a damn."[6]

Amid the jollity, there was a concentrated effort to build a firm base of support in Pierce County. Foster's people sought candidates for the legislature who would promise unyielding support when house and senate met in joint session to vote for senator. They wanted a solid delegation, men with no second choice to whom they might bolt.

The campaign received a setback when in September the Beaver Dam Mill in Cumberland was destroyed by fire. Foster had to rush back to Wisconsin to arrange rebuilding on a smaller scale since the white pine forest was nearly exhausted.

"Would like to be in Tacoma today at Republican county convention," he noted in his journal on October 6, but things could not have gone better had he been on the floor with the delegates at Germania Hall. Every man nominated for the legislature pledged to support him "all the way." On the motion of G. S. Grosscup, an attorney for the

H. G. Foster, who operated a mill of his own in Tacoma, helped in his father's campaign (from Mississippi Valley Lumberman, *May 4, 1894; Minnesota Historical Society*)

Northern Pacific, the convention instructed its nominees "to unitedly use all honorable means to secure Foster's election to the United States Senate."[7]

The *Ledger* ran a two-column picture of the candidate on its front page under the caption "Enthusiastic for Foster." The rival *News* put the convention story on page 3 under the headline:

"IN THE CLUTCHES OF THE OCTOPUS—

REPUBLICAN RING DICTATING EVERY
NOMINATION OF THE CONVENTION—

OPPOSITION RUTHLESSLY SWEPT ASIDE."[8]

Counteracting the complaints about boss politics was the reputation that the St. Paul & Tacoma had won for its role in stabilizing the economy and saving other local institutions during the Panic. The *Mississippi Valley Lumberman*'s account of the senatorial campaign quoted a "West Coast authority" as saying:

During that time when the financial cyclone of '93 swept this and our sister states, young and growing industries were its special victims. During all this period the mill was never shut down; the mines never ceased to work, the coke ovens to burn. Every morning from 10 to 15 hundred men went to work; every month from 10 to 15 hundred men received full pay for a month's labor. Never during this trying time was a pay day missed, never in this time has there been any trouble; never has he known a strike. Not a dollar dividend has been declared on any of these various enterprises, but every dollar earned has gone back into the plant, increasing it, extending its trade, and employing more labor.

The *Puget Sound Lumberman* chimed in on a similar note, recalling the occasion when "a retail lumberman in eastern Washington owed Wheeler, Osgood & Co. and the St. Paul & Tacoma Lumber Company large bills, long overdue. A. G. Foster went over to settle the claims in some way, if possible. When he returned he was asked what he had done, and if he had forced a collection. 'No,' he replied in his cheery way, 'I have never closed a man out yet, and I'm too old to begin now. I told him to go ahead and do the best he could.' Later on, good times came and the accounts were paid in full. No politics in this—just the act of a man with a kindly feeling for his fellow man."[9]

The county convention's instructions to its nominees to support Foster proved no handicap in the November election. Republicans won eleven of the twelve legislative seats in Pierce County. State-wide the party captured both senate and house. Even better for Foster, the Populists did well in King County, thereby weakening the position of his Republican rivals in the area of greatest population.

When the legislature assembled in Olympia in the first week of January 1899, four Republicans dominated talk in the capitol corridors and the after-hours watering holes: Foster, Wilson, Humes, and Levi Ankeny of Walla Walla. Little difference was noted among them concerning issues.

A. E. Henderson, who interviewed Foster for the *Ledger,* condensed the candidate's answers into a single statement:

I think I can do my country good. I would like to be placed in an official position where, with the assistance of others, I could take a hand in the improvement of this country, or, rather, the State of Washington. Our coal fields—there are none better—and our Oriental trade need to be developed. Washington should find a broader market for its coal and from its mines and forests create remunerative labor for thousands more of men. . . .

I shall demand an improvement of our harbors and rivers and coast defenses. I am in favor of a reasonably high tariff and of a gold standard. These will give us stability and resources requisite for needed improvements. I do not care to say what I would first try to accomplish, but this state must have some good public buildings and coast defenses. We must keep the Philippines. I am in favor of liberal appropriations, but a close scrutiny of the expenditures.[10]

This was Republican orthodoxy. There was nothing in the statement that Foster's rivals could not endorse, though Ankeny had for a time flirted with the free silver heresy. The contest would center on geography, personality, personal alliances, and integrity.

Alden Blethen, the choleric editor of the *Seattle Times,* still in a Populist frame of mind, expressed his distaste for the whole lot. Wilson he brushed off as "not a great man, hardly up to the standard of mediocrity as membership in the U.S. Senate goes." Ankeny, the *Times* dismissed as rich, ambitious, and "unqualified by ought save desire." Humes was "a perennial office seeker without merit. An unscrupu-

lous politician, without honesty. A man who would sacrifice every party principle, every honest purpose, for official emoluments . . . without money, without ability, without sufficient education to properly fill the place of Congressman, to say nothing about the position of U.S. Senator, and after the spectacle he made of himself from the standpoint of morality and decency in the campaign that landed him in the mayor's chair ought not to receive a single vote for any office without respect to its character or degree."

After that blast, Blethen's put down of Foster—"eminently respectable because he has demonstrated he can amass a fortune in a legitimate business even though he may have no special qualification for the office he seeks when judged from the standpoint of politicians" but "unthinkable" as senator because Tacoma already was represented by both a congressman of its own, and the state's congressman at large, so that getting a senator would give the city half of the congressional delegation—was almost an endorsement.[11]

Ankeny's candidacy raised the most questions. He was a Walla Walla banker and wheat speculator. During Washington's first sensatorial contest in 1889 he had been suggested as a possibility. The idea got nowhere in the legislature but it burred into Ankeny's consciousness. In subsequent elections he ran steadily and expensively, earning a reputation as a man who told his managers to spend whatever was necessary to get him elected. He was reputed a soft touch for any legislator on the take. The largess of his campaign managers was the stuff of legend.

The *Spokane Outburst* found it "difficult to explain Ankeny's candidacy upon any other ground than money; and if his money is his chief recommendation, how can he dare to attempt to debauch the legislators of this state, and how can any legislator have the effrontery to be a shining example of debauchery?"[12] The *Whatcom Blade* called him "not much of a politician but a good financier" and said that "his negotiations for the purchase of a seat are pursued with the same quiet determination that has characterized the accumulation of the fortune that is now being used to bring that gentleman into prominence."[13]

Ankeny had few legislators openly pledged to his cause. The worry in the Foster camp was that as an eastsider Ankeny would inherit Wilson's supporters if the senator faltered, that the Seattle people backing Humes would never support a Tacoman, and that Ankeny's managers might have purchased a few late-ballot stand-by votes among Democrats and Populists. Under the circumstances, Foster's managers decided to concentrate their attack on Ankeny while being gentle with Wilson in the hope of inheriting his delegation.

When the balloting started in the over-aged wooden building that served as the capitol, Hume, Wilson, and Foster were evenly bunched with votes in the mid-twenties, Ankeny had seventeen, and the minority party vote was scattered. Foster's strength came entirely from the west side of the Cascades, half of it from Pierce County.

Day after day, when their separate sessions ended, sena-tors and representatives gathered to try to fill the senate seat. Day after day no candidate received as many as half of the fifty-seven votes needed for nomination. Day after day the newspapers predicted the imminent collapse of whichever candidate the editor most disliked. The *Tacoma News* had finally hired a new editor, Albert Johnson, who pushed the paper out of the Democratic column and into Foster's camp.

A flamboyant conservative with a lust for combat, Johnson doubled as editor and political reporter. Each afternoon he caught the boat to Olympia to cover the afternoon session; on his way back in the evening he wrote his stories. One day he gave banner-line prominence to the claim of one of Foster's managers that one of Ankeny's managers had tossed five thousand dollars through the transom of the hotel room of a legislator—only to have the money indignantly returned. Johnson took a bundle of the newspapers with him on the steamer to Olympia and went through the hotels where legislators roomed, stuffing copies under every door. No names were named in his story, nothing was even investigated, let alone proved, but Johnson ever afterward insisted that he had prevented a possible stampede toward Ankeny.[14]

After seventeen ballots, no candidate was close to nomination. Then, on January 28 (Foster's sixty-second birthday), two of Wilson's delegates gave up on their man. One went to Ankeny, one to Foster.

Wilson called for a caucus of Republican legislators to see

if there was a candidate the entire delegation would support. He had miscalculated his strength. When it became apparent that he would not be the party favorite, his managers called for all Wilson supporters to walk out, leaving the caucus without a quorum. Two more Wilson people stayed behind. There were no other defections but it was clear that Wilson had passed his peak.

Percy Norton quickly moved behind the scenes. In a conference with Wilson he promised to let him retain control of some patronage in return for his support for Foster. On January 30, the Republicans caucused again. Wilson, "for the good of the party," released his followers and pledged his own vote to the Tacoman. Foster's diary entry that night was terse: "Nominated 8 P.M. by caucus."[15]

When the legislature met in joint session, Foster received all eighty-one of the Republican votes. The Democratic and Populist vote was scattered among seven candidates, none getting more than fifteen.

February 1. Elected U.S. Senator by legislature today—81 votes—only 57 necessary to elect. Overrun by visitors. I made address before legislature. Went to Tacoma. Great reception— brass band—hauled by 4 white horses.[16]

The *Washington Standard* of Olympia, the state's leading Democratic paper, greeted the outcome peevishly:

"Mr. Foster is comparatively unknown in the political world, never having actively engaged in partisan contests. He is a man of family, having a wife, two children, and a grandchild. He has,

likewise, the essential requisite for a member of the American House of Lords—a plethoric barrel. . . . Like most new men in office, he talks freely of what he intends to do when he takes his seat. Among other things, while magnanimously admitting that the official patronage at his disposal does not belong to him, he is equally positive that it belongs to the party which elected him, and he announces that regular machine methods will be observed in ascertaining where it should go. He favors expansion (annexing the Philippines) because he affirms it is the policy of the Administration. A good reason, truly, if it were anyways sure that McKinley has yet made up his mind to 'expand' or not. The remainder of the Republican creed he swallows at a gulp and smacks his lips for more. . . ."[17]

More common were editorials such as one in the *Mississippi Valley Lumberman,* which concluded, "The state of Washington is to be congratulated that it has elected a man of such eminent fitness for this position."[18]

The Fosters entrained for Washington, D.C., on February 12, pausing at St. Paul en route to accept congratulations. Diary extracts reflect the new senator's first day in the capital:

Feb. 24: My credentials presented by Senator J. L. Wilson in open Senate. At capitol all P.M. and on Senate floor
Feb. 27: Called on President at Capitol
March 2. Call on Sec of War, Navy and Agriculture, & at Treasury Dept. on Quartermaster General
March 7. Heartie and I beat Matt and Col. Griggs last night four straight games of whist.[19]

Being senator had not changed Addison Foster.

"The Most Perfect Sawmill"

WHEN THE founding fathers of St. Paul & Tacoma purchased 80,000 acres of Pierce County timberland from the Northern Pacific in 1888, they congratulated themselves on foresight. Ten years later, in a reflective letter to Henry Hewitt, Chauncey Griggs admitted they had misjudged the situation.

"My opinion (and I think yours) is somewhat changed in regard to timberlands in this country," said Griggs. "It is a very different proposition from timberlands in the middle-west." They had been accustomed to yellow pine country "where everybody can log at a very small expense compared with our western country." In the fir, spruce, and cedar forests of the Cascades, "timberland must be in a body or near the waterways to make them valuable, and isolated quarters are almost worthless to the owner unless he can find a purchaser and make arrangements with those who own a large body of land contiguous to the same. When we made this purchase our ideas were somewhat different and we entered into a very large contract which might have resulted badly for us. . . ."[1]

The NP land grant had been in alternate sections, so the St. Paul & Tacoma purchase was too. The problem of blocking up such scattered tracts into a solid expanse economical to log with high-cost steam equipment was compounded by difficulties with the interpretation of the contract, difficulties which though worked out by negotiation might have been avoided had the purchasers been cognizant of the enormity of trees and distances in the Far West. They had created additional difficulties for themselves.

"It was unfortunate that we took hold of so many things without having control or knowing more about them than we really did in investing our money for the purpose of up-building this section of the country," Griggs said. "It was at the same time very unfortunate that you undertook to build up a manufacturing town—the City of Everett. You are well aware I felt very badly in relation to your going ahead so much in that enterprise, and went into more things here to counteract the feeling that was brought out against the company more than anything else. But these things are all in the past. . . ."

The renegotiated contract with the Northern Pacific reduced the cost of bringing logs from the woods and made access to the midwest market at least theoretically profitable. The knowledge of timberland Hewitt had gained as a timber cruiser proved valuable in the development of the company's logging operations. Jones's introduction of the band saw had revolutionized sawmilling in the Douglas fir country, and he continued to introduce new devices. The policy of firming up the tideflats with sawdust and wastewood from the milling process had created forty acres of storage space. Foster's election to the senate provided a spokesman in the national forum for ideas he and his partners believed would benefit company, community, and nation. Among these ideas were a protective tariff, a sound currency, encouragement of trade with the Orient, and harbor improvements by the Corps of Engineers.

"It seems to me that day-light is ahead of us," wrote Griggs to Hewitt, "and that our matters will work out to the interests of those who have invested in our plant as well as in several other concerns that were bad in the first place and which we have resurrected.

"We have a common sympathy between us, and whatever we do should be for the benefit not only of ourselves but those we are interested in. I want to see not only those I am interested in successful, but even the general public that are willing to do anything,—every man should be prospered. Many, we know, will not be because they are of that character and mind that will not be benefitted by any good results. This is really a great misfortune to our country,—in having a class of men who cannot see that it is to their advantage to have others prosperous in the same community."[2]

The panic was indeed over. Business continued to improve. Throughout 1898 and 1899 the mill ran second shifts of ten hours most nights. The cut for 1899 totalled 86 million board feet, 6½ million feet of lath, and 50 million shingles. Shipments went by rail to twenty-three states and territories, and to Mexico, Chile, Australia, China, Japan, England, and South Africa.[3]

Basking in the warmth of prosperity, the board voted unanimously to approve President Griggs's recommendation that the long-delayed second mill be constructed. It was finished in a year and went on the line in December of 1900. A reporter for a midwest trade journal was moved to poetic and philosophic reflection when he came to Tacoma and, from the downtown bluff, looked across the bay:

To the spectator who stands on the cliffs of Tacoma overlooking the waters of Commencement Bay and the long stretch of tide lands across which the waters of the resistless sea ebb and flow in the wonderful operation of the tides, the industrial development centered at the head of the bay forms a picture suggestive of many reflections on the growth of Tacoma and the enterprise of men who have been pioneers in bringing the natural resources of this wealthy country into touch with commerce; waking the slumbering elements and transforming them into active agencies for the creating of business, ministering to the comforts and necessities of man.[4]

The combined capacity of the St. Paul & Tacoma mills was 400,000 feet of lumber and 400,000 shingles in a ten-

hour shift. Mill A had been redesigned to handle Douglas fir exclusively, while in Mill B the largest band saw on the Pacific Coast worked spruce and cedar. A pair of 250-horsepower steam engines, designed and built at the Puget Sound Dry Dock Company plant over in Quartermaster Harbor, drove the machinery in the shingle mill, while in the planing mill nine blocking machines, three surfacers, and three re-saws, each equipped with cut-off saws to trim lumber to exact lengths, finished the product.

The dry kilns were of brick, an innovation in Pacific Coast milling. St. Paul & Tacoma had decided on masonry after losing three wood-walled kilns to fire.

The sheds on reclaimed land could shelter from 3 to 5 million board feet depending on the configuration of the cut. Another 15 to 17 million feet could be stored in open stacks, with almost unlimited space available for more if the need arose. Cut lumber was moved on a system of overhead tramlines that traversed all parts of the yard and extended for a mile down the right-of-way to the ocean dock.[5]

The cargo trade remained an important outlet (and a special room had been fitted out in the headquarters building for the exclusive use of sea captains while their ships were being loaded) but, at last, rail shipment was becoming the dominant factor. Rate cutting begun by James Hill had broadened the market area available. Nearly half of the out-

A turn-of-the-century band saw (from West Coast Lumberman, *1904)*

St. Paul & Tacoma mills A and B with holding ponds (from company letterhead courtesy of Corydon Wagner, Jr.)

put of the two mills was shipped beyond the Rockies.

In 1891, Foster had arranged for some experts from the Chicago, Milwaukee and St. Paul Railway Company to come to Tacoma and test the tensile and lateral strength of Douglas fir. The experiments led to the acceptance of the lumber as standard material for sills and body work in boxcars. The Army Corps of Engineers had approved fir beams for bridge and dam construction. These victories made it easier to introduce Douglas fir as home-building material in areas that previously had thought only of yellow pine.[6]

Three St. Paul & Tacoma logging camps and two camps operated on company land under contract loaded an average of sixty flatcars of logs daily. The great trains clanked down the NP tracks from the foothills of Mount Rainier to the tideflats and spilled their loads into holding ponds adjoining the mills. Jones had devised a system of gates that opened at high tide, then closed at ebb, to keep a constant supply of water. The ponds usually held from 7 to 15 million feet of logs awaiting the glistening bands.

Between five and six hundred men labored daily for St. Paul on the tideflats—and sixty horses pulled wagons that delivered firewood to Tacoma homes, and slabs and edging to manufacturing plants around the tideflats. The monthly payroll at the mills ranged between $30,000 and $35,000. Logging crews pushed the payroll to $50,000.

The entire St. Paul & Tacoma plant site, including the ocean dock, was now lighted by electricity generated by a steam turbine fueled with waste from the mill. On rainless nights Tacomans strolled along Cliff Avenue "to look at the lights on St. Paul." The mill on the boot, like the mountain beyond, was a source of civic pride.

Statistics on mill size and product for the area are uncertain at best, but Tacomans were pleased to note that the owners of the Port Blakely Mill on Bainbridge Island, who had been calling their plant "the largest sawmill in the world," felt obliged to add the phrase "under one roof."

Best of all, for civic boosters, was the description used by the correspondent for the *Mississippi Valley Lumberman:* "The most perfect sawmill."[7]

Organization Man

Seagulls fell dying from the brown and stinking sky. The noonday sun glowed dull behind the murk. Fear hung thick as the pall of smoke over the Pacific Northwest. A myth was going up in flames, the myth that the evergreen forests of the western slope would not burn.

The folk belief in the inflammability of the northwest woods was a strong one. Foresters knew that the sunlight-craving Douglas fir had for thousands of years relied on fire to sweep away the shade-tolerant cedar and hemlock, leaving space for the fir to flourish. Even in the first half-century of the logging era fire consumed more trees in the Cascade Mountains than did saws. But the early fires had been self-contained; they let a little light into the forest, then burned themselves out. In 1902, this was not so.

The summer had been hot and dry, with low humidity. Winds from east of the Cascades moved across the western woods, parching the conifers and turning the usually sponge-like carpet to tinder. In logging camps, it remained business as usual. Clearing and slashing fires were set without regard to air pressure or humidity.

One day in early September when the sky was clear, the sun hot, and rain only a memory, and arid wind funnelling westward through the gap the Columbia River had carved in the mountain barrier fanned small fires burning in slash and undergrowth along the valley of the Lewis River. Sparks from the ground fires rose into the dried needles of the sheltering conifers. The canopy exploded in flame. A firestorm swept the green timber.

The fires converged and moved toward the village of Yacolt, a hamlet of fifteen buildings some twenty miles northeast of Vancouver, Washington. Though the townsfolk all survived by taking refuge in a creek bed and their homes suffered nothing more than heatblisters, the Yacolt Burn has become the generic name for the fires that raged in 700,000 acres of virgin timber between the lower Skagit in northwestern Washington and the upper Willamette in southwest Oregon that month. Sometimes they are called "the Fires of Ought Two."

More than a third of the forest in the area at risk was destroyed—an estimated 12 billion board feet of timber. Thirty-five lives were lost. Included in that statistic were two families with six children who were driving to a picnic when their wagon was overtaken by flames; a rural mailman who lay down by a log, crossed his arms and serenely awaited his fate; a logger who found refuge in a lake only to have a burning tree fall on him; a man who crawled into a hollow log and baked; and a housewife who ran back into a burning cabin to rescue her Singer sewing machine.

When at last the merciful rains came and the smoke cleared and the losses in lives, board feet, and dollars could be reckoned, lumbermen recognized in fire a common enemy and in their individualistic approaches to logging a common enemy. The Yacolt Burn helped bring to an industry that approached anarchy, some order.[1]

Everett Griggs was in every ample inch an organization man. He worked easily with others. He gravitated toward presidencies: of companies, of trade associations, of social clubs. He was elected by fellow officers to the captaincy of his cavalry troop in the National Guard, chosen by superiors to be a major in the regular army. Nowhere was his capacity for organizational leadership more manifest than in the campaign to create some unity among lumbermen on industry-wide projects of mutual benefit.

The first of the second generation among the founding families to go to work for St. Paul & Tacoma, Everett was born in Chaska, Minnesota, on December 27, 1868. He spent his boyhood in the competitive luxury of the Summit Avenue mansions of St. Paul; after graduation from the Sheffield scientific department of Yale University in 1890, he spent a year in Europe, worked another year on construction of the Tacoma, Olympia and Grays Harbor Railroad (which his father had contracted to build for the Northern Pacific), and finally, at the age of twenty-two, went to work in the mill on the tideflats. His first job was as swamper on the green chain (least skilled of jobs), and next as off-bearer for the edging crew (more dog work), but within two years he was assistant superintendent of the mill, and in another year its superintendent.[2]

No one pretended that Everett's rise was simply the result of merit. His father carefully supervised his assignments to make sure he knew in muscle and bone as well as in mind the way the plant operated. But the education was not wasted. For young Griggs, St. Paul & Tacoma was not merely a table of organization, a flow chart and balance sheet, but a process that he could feel, hear, smell. He knew the shudder of the machines, the scream of the thin steel blades, the odor of sawdust. He had exceptional energy. As assistant superintendent and then superintendent it was his pride to be first at the office in the morning—before eight, and last to leave—after six, and never to miss a day. He came back from his honeymoon in Hawaii a week early.[3]

When the board of directors was reorganized in 1900 after the death of Percy Norton, Everett was made second vice-president. (At the same time Henry Hewitt's son John joined the board as assistant treasurer.) In his new capacity,

Everett became the company's liaison man with the industry and its spokesman in dealing with the government.

Almost from the start, northwest lumbermen had dreamed of an organization that might rationalize the chaos of competition by controlling production, standardizing the product, and regulating prices.

Fixed prices had great appeal in an industry with a multitude of small operators and a rigid cost structure, but not even in the most centralized area of the lumber business, shingle-making, did the assumed benefits of an established price list offset the temptation, which for marginal operators often seemed an imperative, to cheat.

The first recorded discussions among Puget Sound lumbermen about grade and price regulation took place at the Tacoma Hotel on October 6, 1891. Percy Norton represented St. Paul & Tacoma at that gathering, and at a subsequent meeting in Seattle where the Northwest Association was formed to set standards for grading lumber, a necessity if Douglas fir was to win acceptance in the Mississippi Valley yards where wholesalers had become accustomed to dealing with graded white pine. The Northwest Association fell apart during the Panic of 1893 when mill owners were less concerned with industry-wide standards than with survival.[4]

In 1895 three Oregon mills formed a pool called the Oregon Pine Company to standardize grades of lumber and administer prices; it held together for almost four years.

Colonel Griggs in 1896 conducted an extensive correspondence and made several trips to California in connection with a proposed Central Lumber Company, which was to bring all the major mills of the entire coast, including British Columbia, into an organization that would regulate grades, prices, and markets. He withdrew St. Paul & Tacoma from participation when he became convinced the proposed company would constitute an illegal trust.

"We were not willing to continue as members unless it was made a company with an invested capital covering the entire interests it represented," he explained in a personal letter to Congressman James Hamilton Lewis. "I have never been in favor of trusts and never acted or intended to act in connection with such organizations unless the entire capital was invested for the purpose of producing manufactures at a less price than could be done otherwise, so that the consumers would have as much benefit as the investors."[5]

The line between a trust and a company was indistinct. A trust meant a business combination by which stockholders of corporations or firms assigned their stock to a small board of trustees with voting or managerial powers and received in return trust certificates on which payments similar to dividends were made from time to time. The purpose of a trust was to bring separate firms into a combination to exercise monopolistic control over a product—which was illegal. The Sherman Anti-trust Act of 1890 forbade "every contract, combination in the form of trust or otherwise, or conspiracy, in restraint of commerce among the several states."

It was the editor of a lumber trade journal who demonstrated that a trade association might achieve the goals of grade and price regulation without running afoul of the law. Victor Beckman had worked on the *Mississippi Valley Lumberman* in St. Paul until 1889, when he moved to Tacoma. Until 1895 he edited the *Puget Sound Timberman,* then became editor and publisher of the *Pacific Lumber Trade Journal.* Beckman won a reputation as the best informed man in the Northwest on the problems of the lumber and shingle trade, and he used every ounce of his influence to persuade mill owners to accept the need for controlled production, uniform grades and prices, and a united front against the railroads.[6]

Freight rates to market areas east of the Rockies were an obsession among Northwest lumbermen. Well they might be. Most of the large mills were owned by men who, like Hewitt, Jones, Griggs, and Foster, had run mills in the Midwest. They knew the upper Mississippi market and knew personally the wholesalers they dealt with, whereas the seaborne trade on which they were forced to rely, was unfamiliar at best and subject to such unpredictable factors as hurricane, revolution, and currency fluctuation. But the Midwest, as St. Paul & Tacoma had found early, could not be penetrated by Douglas fir when the freight rates were sixty cents a hundred pounds. Even at forty cents, to which they fell, transportation costs made up half the retail price on common grades of Douglas fir sold at St. Louis.

In 1900 drought and crop failure in the upper Mississippi area forced farmers to cut back sharply on lumber purchases,

demoralizing the market. Several northwest mill owners approached the railroads with a plea for emergency concessions on rates to Colorado, Nebraska, Kansas, and Missouri. They found themselves stymied by the effective lobbying of associations representing the manufacturers of white and yellow pine lumber, who persuaded the carriers that the requested change would amount to a betrayal of old customers and would cost the railroads business in the long run.[7]

That defeat, reported by Beckman in the *Pacific Lumber Trade Journal* editions of October and November 1900, led Washington lumbermen to form an association of their own. Though member mills were all on Puget Sound or Grays Harbor, they called themselves the Pacific Coast Lumber Manufacturers Association (PCLMA). James E. Bell of the Bell–Nelson Lumber Company in Everett was elected president at the association's first meeting in January of 1901, but Victor Beckman was permanent secretary. Bell–Nelson sold its mill and timber to Weyerhaeuser Timber Company and, in 1902, the PCLMA elected a new president, Everett Griggs.[8]

Everett Griggs' assumption of the presidency of the Pacific Coast Lumber Manufacturers Association coincided with the Yacolt Burn. He turned the disaster of the 1902 fires to some advantage. Though the basic purposes of the PCLMA remained to establish prices and grades for lumber and to win concessions from the railroads, the association responded to the need for fire prevention measures. This not only met

"Going in for the Eastern trade" was the caption for this picture in the Mississippi Valley Lumberman, *March 1, 1901. Chauncey Griggs caresses a log headed for the bandsaw (Minnesota Historical Society)*

a demonstrated need, it was good public relations. Nobody was in favor of forest fires.

The PCLMA encouraged owners of timberland to pool resources for the establishment of lookouts, patrols, and fire-fighting teams in areas where they held adjoining property. George S. Long, manager of the vast holdings that the Weyerhaeuser Timber Company had just purchased out of the NP land grant (a forest-empire ten times the size of St. Paul & Tacoma's), was the catalyst in this campaign. Long paid for a poll conducted by Victor Beckman, which asked lumbermen for suggestions on ways to prevent and limit fires.[9] Among the ideas advanced were a restricted summer season, mandatory shutdowns in periods of low humidity, compulsory slash disposal by approved methods, the stock-piling of fire-fighting equipment at all logging camps, and laws making the owner of timberland fiscally responsible for damage done by a fire that started on his property and spread beyond it. All ideas eventually were adopted.[10]

The Washington legislature responded to the fires of 1902 by creating the posts of state forester and chief fire warden, but the enabling legislation failed to define responsibility or to provide adequate funding.[11] Griggs, Long, Beckman, and others set out to prepare the way for more coherent laws by lobbying and public education.

In 1905 the legislature repealed the act of 1903 and adopted one establishing a state board of forest commissioners empowered to appoint county wardens and deputies as enforcement officers.[12] But again there was a problem with financing. Legislators raised the argument that companies like St. Paul & Tacoma and Weyerhaeuser, which had received, though at second hand, enormous swaths of public land, should not expect the taxpayer to foot the bill for guarding their property against fire. The PCLMA de-fused this issue by backing an amendment that authorized the state fire warden to accept contributions from private interests. The first year, 1906, saw contributions totaling $9,670, with Weyerhaeuser putting up two-fifths, and St. Paul & Tacoma making the second largest payment.[13]

Most of the private fire-fighting groups merged in 1908 into the Washington Forest Fire Association, permitting the coordination of their patrol and firefighting work. The owners of 2½ million acres of forest land agreed to pay two cents an acre annually to supplement the work by state agencies in preventing fires. "These private protective associations spread over Washington, Oregon, California and the Idaho Panhandle between 1905 and 1912," William B. Greeley, head of the Forest Service, was to recall. "They took the lead in combatting forest fires and wrote a deal of forest history. The fire associations had a large part in the development of western fire codes and gave them a practical cast of direct action. Time and again the associations persuaded their state legislatures to pass laws enforcing new techniques or methods of fire control which their own experience had proved efficacious. . . . It was here that industrial forestry began in the West."[14]

Fire prevention work heightened the lumbermen's sense of their stewardship of the forests. The major owners, most of whom had migrated west from areas suffering depletion, knew that forests could disappear, though they were in profound disagreement as to how long that might take. Their stake in the survival of the forests was greater than that of their critics. Though not quick to adopt the proposed programs of the conservationists in their entirety, they did commandeer much of their vocabulary.

"A lumberman is after all a farmer cutting a crop of perpetual evergreen," President Everett Griggs told a meeting of the PCLMA in August of 1905, "and the intelligent application of diversified farming must apply.[15] He expanded on the idea in his address to the annual meeting of the association at the Tacoma Hotel a few months later:

"We are guarding a vast industry representing 196 billion feet of timber, 60 million dollars in payroll, and affecting directly about 100 thousand people. Our eastern competition is fast disappearing. In the east one third of the total output of three states was hemlock in 1904, a wood not considered a factor some years ago. Within five years that vast consumption will begin to draw its main supply from this coast, and with five or six competing transcontinentals and the prospective (Nicaragua) canal I do not hesitate to admonish you to husband your resources. *Slab light, reduce the saw kerf, and keep your eye on the burner.*"[16]

In 1909, the Forest Fire Association changed its name to the Washington Conservation Association. Its officers, who were drawn from the PCLMA, joined with those of a similar group in Oregon to form a broader organization called the Pacific Northwest Forest Protection and Conservation Association. That year the first National Conservation Congress was held in Seattle. The Congress was not devoted entirely to self-congratulation. Senator Albert Beveridge spoke against "the theory that the cutting of the forests conserves the forests." But when the session ended one of the lumbermen spoke of a change in attitude: "Timber owners of the Pacific Northwest are being held up as protectors of the nation's resources instead of destroyers; as worthy of public sympathy and help, rather than of public suspicion. It means conceding to us the honestly earned right to a voice in the law and politics of conservation."[17]

Everett Griggs, president of the PCLMA during the years when the new attitude was formed, pointed out to the 1910 convention why conservation was good for them. "To lessen the supply of an article enhances its value, and therefore conservation means higher prices. With increased stumpage value there will be less waste in logging for a manufacturer would be foolish to throw away any portion of the log that he can sell to advantage."[18]

The campaign for fire protection and conservation was simple compared to the struggle over freight rates. Forest fires had no constituency to defend them, but freight rate regulation had the opposition of the most powerful political interest in the state, the railroad lobby.

A veteran reporter who had covered the state legislature almost from the day of statehood, put it this way: "The state political machine, at times partisan, at other times non-partisan, at still other times by-partisan, has existed ever since the state was admitted in 1889, and has for its chief purpose the enactment of legislation in the interests of the transcontinental railroads and the prevention of legislation for their regulation."[19]

In the same year that Major Griggs became president of the PCLMA, J. F. Farrell took charge of the railroad lobby in Washington on behalf of James J. Hill and the Great Northern. Though bipartisan in politics, Farrell's power was so great that the Republicans once summoned him to a deadlocked caucus to decide who the next governor should be. By playing on natural divisions in the body politic, pitting Seattle against Tacoma, city against farm, eastern Washington against western Washington, Farrell maintained a balance that he could tip in either direction to block legislation that the railroads opposed. When all else failed, minds could be changed by cash alchemy.[20]

Griggs and Farrell fought each other in a classic political duel, their weapons, legislators at close range. The PCLMA began by again asking the carriers for lower rates to the lower Missouri area. Request denied. So the PCLMA carried its fight to Olympia. The state constitution adopted in 1889 authorized the legislature to establish reasonable maximum rates for transportation of both freight and passengers and to create a railroad commission to regulate the carriers. Agricultural interests long had wanted such a commission. Now the lumbermen threw their support behind the idea. Farrell finessed this eastside–westside alliance by making a deal with Levi Ankeny, the perpetual senatorial candidate from Walla Walla. If Ankeny could deliver the votes of his

supporters in the legislature to oppose the creation of a commission, Farrell would throw the railroads' support behind his bid for the senate. Ankeny passed the word to his troops. The commission bill was shunted onto a side track and in 1903 Ankeny at last found himself in Washington, D.C.[21]

Major Griggs learned from that experience. In 1904 all candidates for the legislature were supplied with information on the freight-rate question, including editorials written by Victor Beckman equating lower rates to the Midwest with higher employment in the Far West. The candidates were then interviewed by committees from the PCLMA and asked to pledge their support for any and all bills that could be used to force a lowering of rates. Verbal promises were considered worthless. Candidates had to put their pledges in black and white, duly signed. The names of those committed were then distributed to the 81,000 men who made their living directly or indirectly from the lumber industry.

Out of 136 legislators and state officials elected that November, 108 of them, including Governor A. E. Mead, had signed pledge cards.[22]

The new legislators kept their promises. A bill creating a State Railroad Commission swept through both houses and on March 7, 1905, was signed into law by Governor Mead. The session marked the height of the lumber industry's power in Olympia. The legislators not only created the Commission but dutifully voted down a series of proposals described in the *Pacific Lumber Trade Journal* as "vicious labor measures." These included a "fellow servant" bill to make employers responsible for accidents at work; a bill licensing stationary engineers (which was regarded by management as a scheme to unionize the lumber industry); a bill compelling the payment of wages twice a month and in coin; and a bill fixing the amount to be paid for board in mill and logging-camp company houses.[23]

The lumber industry had established itself as a counterweight to the railroads within Washington State. But the new Railroad Commission could not force the carriers to lower charges on interstate hauls. The fight over freight rates continued.

Aftershocks from the San Francisco earthquake brought significant shifts in relationships between the railroads and the lumber industry in the Far West.

On the morning of April 18, 1906, there was a slippage along the San Andreas fault. San Francisco shuddered and fell apart. The ruins caught fire. For three days and nights a fire storm raged among the ruins. Dynamite used in attempts to create barriers to the flames added to the destruction. Four square miles, five hundred city blocks, were levelled, and three hundred thousand persons were homeless. "Every bank, every theater, every hotel of importance, all newspaper offices, telegraph offices, libraries, municipal buildings, and nearly all the business houses in San Francisco" were gone, according to official report. All that was left was a mocking civic pride that led one San Franciscan

to boast about "the damndest finest ruins, nothing more and nothing less"; pride—and the will to rebuild.[24]

The lumber industry of Puget Sound had been born of San Francisco's need for lumber during the gold rush. Now the industry cranked up again to meet the far greater demand for a new metropolis. During October of 1906 alone, more than 114 million feet of lumber funnelled into the Bay area. Nearly every mill on Puget Sound ran extra shifts. St. Paul & Tacoma double-shifted from April through October. A lumber journal reported that the Weyerhaeuser-owned mill in Everett was running "thirty hours a day." Times had been good before the emergency; now they were great. Demand was up, prices were up, pay was up, and so was the value of timberland. St. Paul & Tacoma holdings in Pierce County increased in estimated value 50 percent during the year.

The inflationary bubble was quick to collapse. San Francisco was deluged with boards and shingles, far more than could be used. The unprecedented movement of lumber to the stricken area disrupted the pattern of boxcar flow on the transcontinental lines. It was all one-way; there was no increase in eastward or northward freight shipments from California. The Bay area was full of empty boxcars.

In the fall of 1906, the Northern Pacific and the Great Northern placed an embargo on western lumber shipments. The Southern Pacific refused to load lumber in Washington or Oregon for shipment south. The Union Pacific would haul lumber but refused to accept Northern Pacific equip-

Company logging railroads had standard-gauge track and were designed for integration into the Northern Pacific system; grade was held to a maximum of 3 percent on the main line. Financial integration with the major haulers was more difficult (from American Lumberman, *May 21, 1921)*

ment. Washington lumber headed for California had to be unloaded from NP cars, then reloaded on UP cars, which increased the cost of the delivered shipments to prohibitive levels.

The quake boom was over, its echo a howl of complaint. Night shifts were dropped. The scream of the saws in Puget Sound mills was replaced by the wail of lumbermen faced with the choice of cutting lumber with no way of getting their product to market, or closing down and being eaten away by insurance and other fixed items of expense.

Everett Griggs and the PCLMA board decided to force the issue. The Hepburn Act of 1906 increased the membership and strengthened the authority of the Interstate Commerce Commission. The PCLMA joined the Shingle Mills Bureau and the Southwest Washington Lumber Manufacturers Association in filing an action with the I.C.C. In March of 1907, in what came to be called the Portland Gateway Case, they asked that the Union Pacific be required to accept Northern Pacific freight cars and equipment on through-shipments, and that the freight rates from western Washington, via Portland, to the mountain and midwestern states be equal to the rate from Portland.[25]

Oregon lumbermen objected to this loss of advantage. The Oregon Association intervened before the I.C.C. on the side of the railroads; the association president, Phillip Buchner, testified that the carriers were managed "by competent men who must look after the railroad interest as we do the lumber interest, and I am glad that no member of this association at this time advocates a forty cent rate to the Missouri river." To Major Griggs and his fellows this was akin to the Pope serving as character witness for Satan. The Washington and Oregon associations severed relations. It seemed that the divide-and-rule technique worked as well interstate as it had in the Washington legislature. But then the railroads overreached.[26]

In June of 1907, with the nation sliding into a recession and the lumber industry of the Northwest already in deep trouble, the carriers announced an increase of ten cents a hundred in freight rates on lumber.

The hike drove the lumbermen back into alliance. Phil Buchner of Oregon wrote Major Griggs of Washington a contrite letter acknowledging past error and promising all possible cooperation in fighting the rate increase. The associations modified their stands on the Gateway issue, Washington toning down its demand for absolute rate equality with Oregon, Oregon dropping its objection to through-shipments of Washington lumber. The PCLMA raised a quick and impressive $100,000 war chest to fight the rate increase. Victor Beckman was put in charge of a joint Washington–Oregon information pool that compiled data documenting the industry's arguments that the new rates amounted to extortion.

The associations won restraining orders from federal courts in both Washington and Oregon suspending the rate increases pending an I.C.C. ruling. Hearings and arguments before the commission stretched from December of 1907 to June of 1908, during which lumber manufacture in the Northwest fell to a ten-year low.

The decision, when it came, was an across-the-board defeat for the railroads. The rate increases were denied. Any payments that had been made before the injunctions were issued had to be returned to the lumber companies. Oregon lumbermen were granted a slight rate differential in the Gateway case but the through-rate was allowed for Washington shipments. The railroads appealed, challenging the constitutionality of the Hepburn Act, but the decision stood.

The lumbermen followed up their victories before the commission and the court by winning a fight in Congress. The interstate commerce law was amended to permit the Commission to review proposed freight rate increases before they became effective. Major Griggs hailed the triumphs as "a great thing from the fact that we all stood together. The dynamic force of associated concentrated effort will jar even a railroad man."[27]

While fighting to make the railroads lower their freight rates, the PCLMA never wavered in its belief that the price for its own products should be set by the industry and maintained at a high level. The united front that prevailed against a common enemy, however, could not be main-

Red cedar siding, knot-free and durable, was one of the company's favorite products (from West Coast Lumberman)

tained within the industry with regard to standards and prices for lumber.

When first organized the Pacific Coast Lumber Manufacturers Association acted as if the Sherman Anti-trust Act did not apply to the lumber business. Price lists were drawn up by standing committees composed of major Puget Sound manufacturers, then submitted to a general meeting of the Association for ratification. When accepted they were published. Efforts were made to coordinate the PCLMA prices with those of the Oregon Association and the Southwest Washington Lumber Manufacturers Association, though discrepancies remained.

Enforcement was largely a matter of persuasion and peer pressure. Mill owners suspected of undercutting on prices or varying on grade standards were asked by letter to explain. A statement that the lower price was the result of misinterpretation or a bookkeeping error was usually accepted—the first time. Persistent price cutters were visited by a delegation of other lumbermen and their derelictions publicized within the association. Sometimes bankers were persuaded to advise their clients to shape up.

In 1906, responding to increased activity by the Justice Department against trade associations across the nation, the PCLMA stopped printing price lists under its own name. Instead, Victor Beckman gathered price figures twice a month from the mills and published lists of "prevailing prices" under private auspices. Enforcement procedures became less blunt, but J. H. Bloedel of Bellingham told a meeting of the National Lumber Manufacturers Association in 1910 that "if there is a weak sister in the bunch he gets his backbone stiffened."

There were always "weak sisters" who found it to their advantage to cut prices. Success in maintaining agreed-upon price levels depended on the state of the economy. In good times, when demand was high, there was little price-cutting. In bad times, when even the strong had to scramble to survive, the lists became, in Griggs's phrase, "little more than a very satisfactory piece of paper . . . punctured from stem to stern."[28]

The U.S. Bureau of Corporations, which issued a massive four-volume study of the lumber industry in 1913, contended that the associations had met with some success in their attempts to set prices. "They have made prices obtained by consumers higher than they would have paid had such activities not taken place," the report stated. But the Bureau conceded that in general the rise in lumber prices stemmed from external factors such as population growth, periods of general prosperity, and "the increasing demand for lumber in the face of a rapidly diminishing ultimate supply of standing timber."[29]

As for Everett Griggs, he found his efforts at stabilizing prices less than a success. "Everybody seems to be fighting everybody else," he said in 1911, "yet if we try to get together and secure better prices, we may go to jail. I don't know but what I would rather go to jail than continue in the lumber business under the present conditions."[30]

He neither quit nor went to jail. Instead he took the lead in pulling the three major northwest associations, the PCLMA, the Southwest Washington Lumber Manufacturers Association, and the Oregon Association, into a new body known as the West Coast Lumber Manufacturers Association. He was, of course, elected its first president.[31]

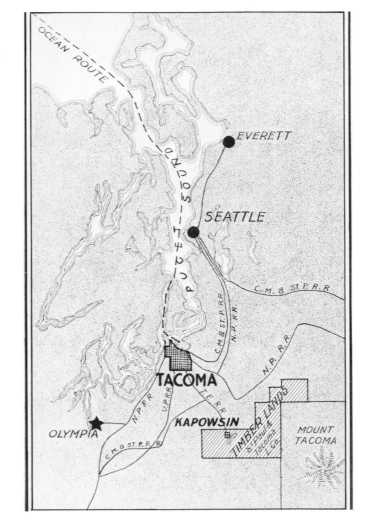

Map showing the location of timber lands of the St. Paul & Tacoma and the rail and water routes from the mills (from American Lumberman, *May 21, 1921)*

The Changing of the Guard

I   N FEBRUARY 1908, Chauncey Griggs, the oldest of the
    founding fathers of St. Paul & Tacoma and its president
    since the company's incorporation twenty years before,
stepped down in favor of his son Everett. The elder Griggs
remained chairman of the board, visited the office regularly,
and was consulted on policy decisions, but he was seventy-
five years old and the flame burned low.[1]

Though bulky in the tradition of successful businessmen
of the Edwardian era, the Colonel had never been robust
following his civil War experiences. He suffered intermit-
tently from malaria, had frequent kidney infections, and in
later years had occasional dizzy spells. His son-in-law the
doctor, George Wagner, tried to persuade him to give up
horseback riding lest he faint and fall from the saddle.
Griggs merely added to the doctor's alarm by having his an-
kles tied to the stirrups when he went for a canter in
Wright Park.[2]

Nor would he stick to his diet. He was more accustomed
to giving orders than taking them. He liked to visit the
Union Club for a quick hello, which translated as "chocolate
cake," and a hand of whist, which meant "apple pie a la

mode." But after two heart attacks and a fainting spell in
the bathtub that led one paper to run a premature obituary,
the Colonel became a stay-at-home, seldom leaving the
ground floor of the big house at North Fourth and Tacoma
Avenue.

Hewitt, who still lived next door at Fourth and E,
stumped in nearly every day for a chat. At sixty-nine, Hewitt was still in vigorous pursuit of opportunity and liked to
talk to Chauncey of new ventures in which he was involved—of the Cordova Copper Company in Alaska, which
he had organized within the year and which was already delivering ore to the Tacoma smelter, now the property of the
Guggenheims; of his plans for an irrigated ranch of six
thousand acres near Lakeview in eastern Oregon; of a new
city on Coos Bay, which he hoped to make the greatest port
between San Francisco and Portland; and of the interurban
route connecting the towns of Marshfield and North Bend.[3]

Time had worn some of the rough edges off Hewitt, time
and success. He no longer made jokes about brushing his
teeth with his finger, nor did he try to embarrass his
daughters, Clara and Mary, by eating peas off his knife just

because they had brought friends home for dinner. Not that he minded being thought old-fashioned. He had been the first of the founding fathers to buy an automobile but his misadventures at driving were monumental and now he proudly drove a buggy.

He affected dark broadcloth suits, and high celluoid collars with stylish wings, but he was happy when he could slip into old clothes and go off to cruise timber or study mineral areas and play the rube while negotiating to buy land by the township. Some said he owned more timberland than anyone in the world except Czar Alexander of Russia; Henry liked to boast that his personal holdings of standing timber exceeded those of Frederick Weyerhaeuser. He was thought to be the wealthiest of the St. Paul & Tacoma partners. Once when Chauncey was teasing him about retiring a soup-stained suit, he answered, "Colonel, I can buy a new suit anytime I want and have money left to buy you out."

Like Griggs, he had become a major contributor to Tacoma causes, especially the First Congregational Church. Neither was a communicant, but their wives were. Once when asked what the four turrets on the church tower represented, Hewitt replied "These are the Father, the Son, the Holy Ghost, and the St. Paul & Tacoma Lumber Company." He and Griggs each gave $10,000 to make possible the YMCA building at St. Helens and Fawcett Avenue, and each contributed substantially to the College of Puget Sound. Hewitt had a special interest in the Ferry Museum of the Washington State Historical Society and, with Robert Laird McCormick of the Weyerhaeuser Timber Company, was the principal contributor to its construction.[4]

The greatest bond between Hewitt and Griggs was not their business association but their interest in each other's sons. The Colonel liked to talk about C. Milton and Theodore, both active in the family wholesale grocery business in St. Paul, Herbert Stanton, the attorney, and Major Everett in Tacoma. All four were graduates of Yale, as were two of the three Hewitt sons. Will, the eldest Hewitt, was a lumberman in his own right, soon to be associated with the management of St. Paul & Tacoma; Henry was helping his father keep track of the many family enterprises, and John had become, in his father's estimate, the best timber cruiser he had ever known and the most precise evaluator of coal deposits.

C. H. Jones called in to see the Colonel, too, but less frequently. He had purchased controlling interest in the Northwestern Lumber Company mill in Hoquiam, and the special problems of Gray's Harbor logging and sawmilling—verticle terrain, huge logs, and a rebellious work force—kept him away from the Jones house at 323 North Fifth much of the time. C. H. was leaner now than ever, his dark hair shot with iron gray, his natural solemnness deepened by the death of his beloved adopted daughter. What talk he had centered on machinery, prices, and politics. It was politics that made him raise his voice. The country was going to hell in a handbasket, what with anar-

The younger members of the Griggs family assembled in April, 1909, for the golden wedding anniversary of Chauncey and Martha Ann. Left to right, seated: Heartie, Herbert, Milton, and Anna; standing: Everett and Theodore (courtesy of Chauncey Leavenworth Griggs)

chists and Wobblies, socialists and unions and taxes, faster now, all the time faster.[5]

The most relaxing visits for Chauncey were those with Add Foster, his arrival always heralded by his rolling laugh and wheezy breathing; like the Colonel, the Senator was not given to leanness. Sometimes he brought his old Airedale, Bam, who was famous for drinking coffee from the fingerbowls.

The old partners still talked some business; they were closing down the Beaver Dam Mill—Wisconsin's white pine was gone, and they still had some land to sell in Dakota. They laughed over their mine in Montana that "almost had gold," and argued gently about whether it was better not to have been a United States senator, each favoring the other's fate. Foster wished he had been able to push through his bills to improve the Tacoma harbor; he didn't mind not having a second term but he wished people remembered that he had introduced more bills that passed (46) than had all four previous senators put together (37). Griggs wished he had been able to save the Crescent Creamery from bankruptcy and speculated on how rich he would have been if he had taken that Eastman stock from Henry Strong instead of $10,000 cash.

The old men knew each other so well that they told each other's family stories. Griggs would remind Foster of the time he'd taken his newborn son for a walk and kept him out so long that the baby began to cry from hunger. "So you just walked up to the door of the nearest house, knocked, asked if there was a nursing mother in the place, and handed little Harry to her. Oh, Martha wasn't pleased with you when she learned." And Foster would remind Griggs of the time he protected his wife and children from a dead mouse and swatted it so hard it splashed on his new coat.

One or the other would recall the time they ordered such a large shipment of tea from Canton for use in the logging camps and sale in the company store that the Chinese merchant had a special label designed for the tea chests. It bore the name "Lumberman's Delight" over a sketch of Mill A against a background of forest and the Mountain. A handsome picture, they'd thought, until they noticed the artist's version of the St. Paul & Tacoma transportation system. Down from the Mountain came a line of Chinese, each tenderly cradling a log in his arms.

There was little need for them to ask about each other's children. The kids had grown up together. All three of the Foster boys were working for the company; Harry, the firstborn (he of the impromptu wet nurse), was in charge of sales in Chicago, Arthur and Everett were stationed in Tacoma. Harry's wife Elizabeth was a beauty, the queen of Tacoma's first Rose Festival in 1896, crowned by Mayor Angelo Fawcett in a ceremony in Wright Park on the Fourth of July.

The following year Griggs's daughter Anna had presided over the festival, leading the dances at the Exposition Hall (which had since burned) and receiving a coronet of roses from a La France bush planted in Tacoma in 1878 when the population was less than a thousand. Foster remembered that the coronation took place in what the papers particularly called "a heavy dew that came dangerously near being a drizzling rain."

Griggs and Foster played whist and dominoes, shared an occasional milk punch, deplored politics. Often they just sat looking out at the bay and thinking of the changes till the Mountain turned pink in reflected sunset, then faded to a memory of steel gray lines against whiteness. There was much to be proud of but their days of accomplishment were over.

When the Tacoma and St. Paul branches of the Griggs family gathered in the big house in April of 1909 to celebrate the fiftieth wedding anniversary of Chauncey and Martha Ann it was with the realization that the next full family gathering would likely be a funeral. Still it was not solemn. After the family dinner and speeches, Herbert, the attorney, read a poem of his own composition in which the spirit of fun, affection, and pride overcame the infelicities of meter and rhyme.

After acknowledging

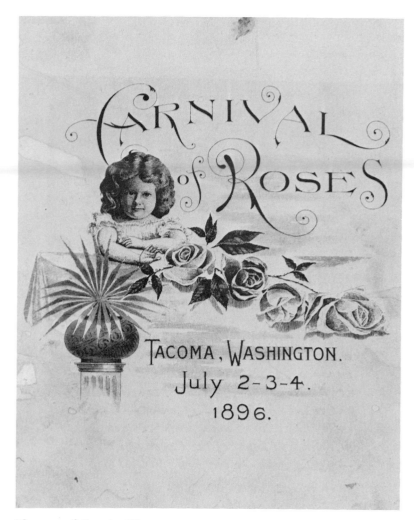

(*Courtesy of Corydon Wagner, Jr.*)

> We're mighty glad
> That ma and dad
>   Married into our family
> And got busy making money

Herbert told of the colonel's battle with the mouse, of the time he fell from the carriage

> And then for sentiment
> Applies strong horse liniment

and of the time when Anna

> Dressed in pa's boots, cane and hat
> Came in saying "Mat!
> What have you been buying today?"

In conclusion

> We confess we're right proud of our dad
> For he fought the fight
> With all of his might
> That the good in man fights against bad.[6]

Colonel Griggs died quietly in his sleep on October 29, 1910. The end came in his bedroom overlooking the bay, the mill he had helped create, and the city he had seen through the Panic of 1893. Beyond the mountains lay the memories of his earlier careers in fuel, steamboating, groceries, and real estate, the trauma of the years of fraternal war, the rigors of a New England upbringing. His had been a life of much accomplishment. He was a man of many virtues, the greatest of which was integrity.[7]

Martha Ann died three years later. Her eldest son, Chauncey Milton, wrote a long letter to his own sons about her life of seventy-eight years, and of what lessons he drew from his parents' example:

At the end of 30 years of life in Minnesota, they had saved a fortune of one million dollars and raised six children. Her ability had made this possible, just as much as his. And through it all we lived well. We did not worry because we did not "have it all."

This grand woman was a pillar of the Plymouth Congregation Church, the President of the Protestant Orphan Asylum, kept up her oil painting, did some of her music, studied German, ran the house, drove the horses, if necessary would fill the furnace if the man didn't get around on time to suit her. Planned for her children's going to school, insisted on their going to college, taking music lessons, language lessons, had much more enthusiasm than any of her children for study and research. . . .

Took the four youngest of her children to Europe and lived in Germany so as to get for them and herself a better knowledge of language, of art and music. And she lived within her income doing it, within the amount stipulated by her husband as his being able to spend on such an undertaking.

And then after all this and getting into a new and beautiful home on Summit Avenue . . . pulling up stakes at the age of 54 and going again with her husband, pioneering in a sense, way out here to the Pacific Coast and rounding out her career there. . . .

My dear boys, it's doing good, it's doing best in one's everyday life that counts. It counts more to pay one's bills and save one's capital first. That is the foundation for every bit of strength one has,—to live within one's means, one's income, to live without jealousy, to be reasonably independent, to be fair and charitable and industrious.

That's the kind of life that is inspiring in the end. The socialists and the spendthrifts and some of the muckrakers and fakirs are trying to discountenance and belittle that kind of life and make the people think and believe that prudence and foresight and thriftiness are synonymous with greed and snobbery and trickery.

What a mass of rot and wicked doctrine has been let loose on the public of late years! It's health and industry and acquisition of capital (not destruction of capital) which has made all progress in this world, spiritual as well as material.

Here was a woman who did her share, not only in the acquisition of her family's capital but in helping others to help themselves.

You boys have a grand inheritance from your immediate grand-parents on both sides. They were producers, they were workers, and after all it's the only kind of life that brings satisfac-tion. . . .

I remember hearing the finest hunting dog breeder in this coun-try, Olf Watson, saying to a party who was telling him of all the good points of a certain young dog, how he'd point and pack, what style he had, how beautiful he was, etc., "Will he *run,* Bobbie? Will he run?" It's just so with man. No matter so much what his graces, his accomplishments, his style, or even his tal-ents and abilities, it's after all—
  "Will he work, Bobbie? Will he work?"

FATHER [8]
Tacoma, Washington
April 16, 1913

Disaster and Discord

NINETEEN TWELVE WAS A good year for Puget
Sound lumber mills. Demand was high and stumpage available. St. Paul & Tacoma's Mill B, the
new one, was double-shifting, turning out lumber for two
ten-hour shifts, six days a week. Mill A, the original plant,
ran only a day shift.

On Friday, June 7, a problem developed with the electric
motor for the re-saw in the planing mill. It would run but
the commutator emitted a shower of sparks. Mill Superintendent H. W. Palmer visited the room housing the motor
several times but could not determine what was wrong. He
did not feel it necessary to shut down the re-saw before the
end of the regular shift.

Shortly before dawn on Saturday, the night watchman,
Edward Pielow, opened the door to the planing mill and
saw a curtain of flame, five to six feet high, leaping from
the shavings and sawdust on the floor. (Later it was theo-
rized that a spark had fallen in debris, perhaps on an oily
rag, where it glowed and smoldered until a strong wind
that rose during the night stirred the air inside the build-

ing, giving life to the blaze that burst through the tinder
dry shavings and powdery sawdust.) Shouting for help, Pielow ran outside to the nearest hydrant, turned on the fire-
hose, and went back to the mill.

Flames had spread completely across the back half of the
building. The stream from the hose made small rents in the
wall of fire, holes that closed the moment the water was directed elsewhere. The flames kept spreading. Heat and
smoke drove Pielow outside. Windows exploded outward.
He played water through the windows and onto the roof but
the fire broke through. In moments the mill was totally involved, a rectangle of flame from which a pillar of smoke
and sparks rose, the orange fading to gray as the plume
spread. Glowing shards of the roofboards rained on the surrounding buildings.

Someone at Mill B had heard Pielow's cry for help. Work
crews of the night shift poured from the buildings and
manned the hoses, but in the excitement no one blew the
fire whistle to alert the town. Not until a pedestrian on
Stadium Way saw the flames was the fire department noti-

fied. Three companies were on the scene before the mill's alarm whistle sounded. Chief George McAlvey sent for all available equipment. Nine fire companies and three engine companies converged on Mill A, too late.

May and June had been almost rainless. Water pressure was low, the wooden buildings dry as kindling. A strong wind from the southwest swept oxygen to the flames that raced along the sawdust sprinkled earth nearly as fast as a man could run. The fire companies gave up on the planing mill but fought to save the rest of the plant. They ran lines into the mill pond and used the steam pump in the fire room of Mill A to throw protective streams on the other buildings, but the roof of Mill A itself, big as a football field, long the pride of the tideflats, flared in a score of places. The mill burned from the top down. The fire room below exploded. The pump melted.

Two storage sheds, the dry kiln, 250,000 board feet of stock lumber, the ties of the railroad track—all were aflame. The very ground itself, firmed up as it had been with a quarter-century deposit of sawdust, caught fire. An area a quarter-mile square was a single glare on the dark expanse of tideflat and bay. Water poured into the heat vaporized without touching anything solid. Paint blistered on the plant's boarding house more than a hundred yards away. Burning shingles drifted out of the sky onto the barn that housed more than a hundred work-horses used for delivering cut lumber and fuel. Fire bracketed three sides of the barn, and firemen concentrated the streams from a dozen hoses on the remaining wall as volunteers went into the smoke and led the horses, one at a time, to safety. All were rescued.

Mill B, the new plant, was saved, thanks as much to the direction of the wind as to the efforts of the firemen. And the southwester, rare for June, kept sparks from crossing the city waterway into downtown Tacoma.

The summer sun rose on a ruined mill. One smokestack, held by its cables, leaned grotesquely over the embers. Charred timbers of building frames pointed to where the roofs had been. Warped and twisted machinery glowed sullenly. Ten hours after the fire started the earth hissed as damp-down streams fell on it. The dead smell of watered ashes hung over Tacoma.

Major Griggs estimated the loss at half a million dollars, all insured. Rebuilding, he said, would start at once.[1]

The death of Chauncey Griggs and the destruction of Mill A precipitated the first serious struggle for control of St. Paul & Tacoma policy. There had, of course, been differences among the officers before, but the Colonel, as the oldest and most experienced of the founders, had been the natural choice for president. On important matters his ideas usually prevailed at board meetings. His emergency decisions were seldom questioned.

Griggs had groomed his son Everett as his successor. Foster and Hewitt, who each also had sons experienced in the trade and serving on the board, did not object when Everett served as acting president in Chauncey's absence. When

Chauncey retired, Everett was the unanimous choice for president and, after the Colonel's death, became chairman of the board.

The first of the company officers to have a training in science, Major Griggs was eager to adopt new methods, to put his own stamp on operations. Even before he became president he had urged the building of a new planing mill and the conversion of the St. Paul & Tacoma plant from steam power to electricity.

After the fire, Everett, without consulting his board, ordered construction of a new planing plant. It was huge, 400-by-250 feet, and its equipment was to be driven by electricity.[2]

The board muttered but did not rebel. Then Everett presented them with the idea of building a new two-band sawmill capable of cutting 250,000 feet per ten-hour day and a new powerhouse to supply electricity for the entire plant. He had plans prepared by W. A. Wilkinson and Son of Milwaukee, who a few years before had designed a modern mill for the Potlatch Lumber Company in Idaho.[3]

The proposal reflected an optimistic view of the lumber market in the years ahead. Though 1910 and 1911 had been poor years, with prices averaging $11.05 per thousand and sometimes falling below the cost of production, there had been improvement in 1912, with an average price of $11.58. The Panama Canal was expected to open new markets. The West Coast Lumbermen's Association and the National Lumber Manufacturers Association, of which Griggs

E.G. GRIGGS
PRESIDENT

Everett Griggs, second president of the company, found himself in the position of a young general commanding far more experienced colonels (*from* American Lumberman, *May 21, 1921*)

was president in 1911 and 1912, had won concessions on tariff legislation, had turned the conservation movement in paths favorable to large timberholders, and had won passage of industrial insurance laws that limited accident litigation. The Washington Railroad Commission favored lower freight rates. Population in the Northwest continued to expand and the value of Douglas fir timberland to increase; standing timber had gone up from an average of 20 cents a thousand board feet in 1900 to $1.50 in 1913.[4] To the Major this seemed the time to invest in plant, even if it meant borrowing.

The old-timers were outraged. Everett Griggs was forty-five years old; Hewitt was seventy-three, Jones, sixty-eight, and Foster, seventy-eight. The elders had doubts about the market, doubts about electricity, doubts about the new machinery, but no doubt at all that young Griggs was preempting their right of decision on matters of fundamental policy. They interpreted the facts he cited as leading to an opposite conclusion. Hewitt said that the increased value of standing timber, along with the rising cost of labor, indicated that it would be more profitable for St. Paul & Tacoma to sell logs for a sure profit instead of risking the vicissitudes of manufacturing them into lumber. Foster doubted the wisdom of going into debt to increase plant capacity under existing circumstances. Jones, who was having both labor and market problems at Hoquiam, objected not only to what Griggs proposed but to what he had done.[5]

Griggs brought the matter to a head by writing a formal letter to the board on February 18, 1913, proposing that work begin immediately on "a two-band mill capable of cutting 250,000 feet a day of ten hours" and on a new powerhouse. He made some concessions. The frame would be built for the entire mill but only enough equipment would be installed to operate one side—"a single-band re-saw, slab slashers, edger, automatic trimmer and lath mill with all necessary rolls and sorting chains, and a turbo generator." This he estimated would cost $150,400 for the mill (of which $21,800 would be for lumber already at hand), and $62,584 for the powerhouse, making a total cash outlay of $200,113.[6]

Jones wrote back with crusty indignation:

Your favor of the 18th received. Note what you say in regard to the "difference of our opinions . . ."; also that you would like for me to make a trip to the Potlatch Lumber Company's plant at Elk River, Idaho, and if possible have Mr. Hewitt make this trip and see Mr. Deary.

I cannot take the time right now to be away; besides do not think a mill cutting small and short logs is the kind of plant to visit to get the best ideas for operating here on the coast. We know that you must have given a good deal of touring and time in making investigations of different mills, but I think you have gotten the idea that the Company need too large and expensive a mill, as the plans you have had Wilkinson draw are altogether too expensive and elaborate for us to adopt and I am sincere when I say that there should not be any such mill built, and that it was all wrong to get such plans and incur such expense before bringing it up with your Board of Directors also the building of a new boom and the building of the new planing mill building at this time, and wrecking the part of the old mill that did not burn and preparing the grounds at so much expense before anything was brought before the directors, and some definite decision arrived at. And I do not think it right or good business to go ahead with a large great outlay for a new mill, even if you have already paid out so much in getting ready.

The past few years' business has shown the profits were not in the manufacturing and operating of the mills, but from other lines and the logging operation. And I truly believe it best to only operate the mill we now have and get everything working on the most economical and harmonious lines. If we produce more logs than the mill can manufacture there is a good market to sell in, and surely at a good profit; and I doubt the profit that will be made in manufacturing. And if a time comes when there can be more profit shown by operating another mill, then at that time let us build one. From the results that are shown in the Creosoting plant it would be much better to put money into that than into a mill at present, and I truly believe that the best thing for our Company, is to put the other engine—that is the present electric plant into condition to operate, and a new dynamo and install sufficient boilers to operate this electric plant to its full capacity, which should furnish all electricity needed, supply all necessary steam for the creosoting plant, when enlarged, and everybody work in harmony for the best interest of the St. Paul & Tacoma Lumber Company and put her on a dividend paying basis, before expending any large amounts on any more sawmills.

This is my opinion of what the St. Paul & Tacoma Lumber Company should do. Later, when shown that more mill capacity will be for the best good and plenty of funds have been accumulated,

then build the mill that is shown to be wanted when the time comes.

> Very truly yours,
> C H Jones[7]

Griggs replied that Jones "did not seem to give me credit for the ordinary intelligence which an operator must have." He said that he "appreciated the character of the timber in which the Elk River plant was operating, but the application of electricity and its operation can be as well studied there as elsewhere." Further, and testily, he said, "If you will follow our own operations, you will appreciate that a large percent . . . are now in timber which is small and which has got to be handled economically in order to be handled profitably."

He claimed that the construction of a new planing mill had previously been authorized and should have been built long ago. All that had been done as to wrecking the old mill and preparing the ground was "what was absolutely necessary to clear the ground of menacing operations." As for the replaced tracks, they were on ties that had rotted out. As for selling the logs, "certainly if other manufacturers can make money and pay the going prices for logs, we should be able to do so. . . . I am convinced that our operations both in the woods and in the manufacturing line can best be conducted by equipping ourselves right electrically, and with the idea of further extensions."[8]

The board had appointed an ad hoc committee to review "the question of building a new sawmill and the type of mill to be built." The committee was composed of Jones and H. G. Foster, Addison's son who had taken his father's place on the board. Its recommendation was to postpone construction indefinitely, largely because the proposed mill could not be erected without borrowing money.[9]

In a supplemental report, Harry Foster said that the 1888 timber purchase gave St. Paul & Tacoma "a heritage second to none on the Pacific Coast, one which "after twenty-four years is many times more valuable than when purchased." But, he reminded his fellow board members, "in our contract with the Northern Pacific Railway, it is necessary for us to operate two mills. Mill A was burned and now it is incumbent upon us to rebuild." After expressing his disapproval of "borrowing large amounts of paper for the purpose of extensive improvements . . . with labor troubles staring us in the face and a new administration [President Wilson] beginning," Foster went on to say,

. . . the kind of mill matters very little so long as all pull together and will work to make a success financially. Every company has internal differences but it should be like a man and his wife, namely: fight it out at home and pull together. No company can be a success without a Captain or leader and he should lead, giving confidence to all and placing confidence in his associates and employees. Team work is what counts. Responsibility should be placed and authority go with it or there is no confidence. Confidence will make success and there is no excuse for the St. Paul & Tacoma Lumber Company not making large returns for the money invested.

What we need more than anything else is an established policy, defining responsibility and authority. An executive Board of Three should work together on all matters and report to the Board of Trustees. The executive Board should have the absolute management of the Company's affairs under the leadership of its Chairman. Each department of this Company should have its head, with confidence placed in that head, with authority. If he cannot make good get someone who will. Get together. Now is the time to make money and satisfy all.[10]

Peace was not to be made with a pep talk. In advance of the next annual meeting, the opposing factions sought proxies among the 15,000 shares. The issue over which the battle was joined was a proposal submitted by John Hewitt, Henry's son, "on behalf of parties in interest with regard to leasing the property of the St. Paul & Tacoma Lumber Company and operating same, cut and handle the timber to be furnished by the St. Paul & Tacoma Lumber Company." Acceptance of the offer would have put the Hewitt–Jones group in charge.[11]

The question of the lease was not submitted directly to the shareholders. The issue of control centered on a vote, sponsored by the Griggs group, to reduce the board of trustees to seven members, from ten. When the meeting convened in the company office on the tideflats, only 25 of the 15,000 voting shares were not represented in person or in proxy.[12]

C. W. Griggs Investment Company (Major Griggs, proxy) held the largest block, 3,736 shares; Henry Hewitt, Jr., (John Hewitt, proxy), the second largest, 3,498⅞ shares. Every relative and in-law, every friend and trust account, had been solicited for support. The biggest undecided blocks were believed to belong to the Browne family and the Howarth brothers.

George Browne had died in 1912, but his eldest son, George A. Browne was present with proxies for 1,800 votes. He cast them with his father's generation, Hewitt and Jones, against the Griggses. The Howarths were in their fifties—younger than Hewitt and Jones, older than Major Griggs.[13] They had come to the Northwest in connection with the Hewitt enterprises. William, now manager of the Everett Pulp and Paper Company, had a stockholder's relationship with St. Paul & Tacoma. Leonard, the lonely and lame bachelor, had been with the Tacoma company since 1891. Though he had worked directly for Henry Hewitt in Wisconsin, he had become the alter ego of Foster and Griggs during the long receivership struggles after the Panic. His must have been the hardest choice. The Howarths' 226 votes went to Griggs.

The Griggs motion approving the smaller board won, 7,890½ to 7,084½. (Years later it came to be said that Leonard Howarth had decided the issue. With the Griggses winning by 806 votes and the Howarths providing only 226 of them, this could not have been the case.)

A new board of seven members was elected. Herbert Griggs placed in nomination the names of himself, his brother Everett, his brother C. Milton, William Howarth, Henry Hewitt, John Hewitt, and C. H. Jones. Jones seconded the nominations of five of the seven—himself, both Hewitts, and Herbert and Everett Griggs. He then moved that these five be voted on first. They were unanimously approved.

William S. Shank then seconded the nomination of William Howarth and C. M. Griggs as permanent trustees (which would give the Griggs group a four to three margin on the seven-man board) and also proposed the nominations of Leonard Howarth, C. A. Foster, and himself as temporary trustees to serve only until the secretary of Washington State formally approved the reduction in size of the board. His motion was adopted, 8,015½ to 6,959½.[14]

Major Griggs had fought off the Hewitts' bid for power but had not established command. At the next board meeting, he was re-elected president, with John Hewitt and C. H. Jones voting nay. Henry Hewitt was elected chairman of the board. Nothing was said about building the new mill. It was not built for years. By then the world and the industry was much changed.[15]

St. Paul & Tacoma was not alone in considering a major mill building project in 1913. The Weyerhaeuser Timber

Company, too, was studying the merits of a modern mill based on advances in electrical technology. Weyerhaeuser, which held far more timberland than any other company, had cut little of it. Their only mill was the second-hand plant they had bought in Everett, rebuilt, and run as a way of becoming familiar with Puget Sound problems.

The mill showed a profit of $194,799 in 1913, but some of the Weyerhaeuser directors, like their counterparts at St. Paul & Tacoma, doubted that the time had come to start cutting on a grand scale. On the other hand the prospects of new markets beyond the Panama Canal, the pressure created by charges in the muckraking monthlies that Weyerhaeuser was in the real estate business rather than lumber, and the insistence by George Long that the time was right, led to authorization of a new mill at Everett.

Weyerhaeuser's Mill B opened in April, 1915: a double band-saw operation, all electric, in a plant four hundred feet long. "The size of the new plant . . . rather staggers me," F. E. Weyerhaeuser wrote to a friend. But for its first half-year, Mill B showed a profit of $163,000. For all of 1916, $606,007.

George Long, with characteristic understatement, summed it up: "It certainly looks like it paid to build the new mill."[16]

St. Paul & Tacoma at War

NOTHING WENT as anticipated in the years immediately following 1913. The Panama Canal, that long-awaited, much-delayed gateway to the markets of the Atlantic rim opened in August of 1914. So did the First World War. Security regulations imposed before the United States entered the war limited the use of the canal for civilian shipments. After the long wait for the canal to be completed came even more tantalizing years of awaiting permission to send lumber through it.

Rail construction lagged. Not only were the carriers buying no lumber for their own use, their rates still did not permit Douglas fir and red cedar to penetrate the East Coast in sufficient quantity to absorb the increased productive capacity of Pacific Northwest mills. Throughout 1914 and 1915 orders declined.[1] The big mill on the tideflats ran at less than half capacity and even so the stacks of finished boards taxed storage capacity.

To bolster its faltering cargo trade, St. Paul & Tacoma joined the Douglas Fir Exploitation and Export Company which had been formed earlier by lumber manufacturers in northern California. The DFE&E, which had headquarters in San Francisco, promoted sales in countries unfamiliar with tall trees, let alone the special virtues of Douglas fir. Samples were sent to dealers around the world, and potential large-scale customers were freeloaded to the States to visit showplace mills such as the St. Paul & Tacoma. It worked. The export trade began to increase. When the United States declared war on Germany in 1917, American ships became legitimate prey to U-boats, orders from abroad plunged, and the DFE&E lost half its membership, though St. Paul & Tacoma was one of the seventeen companies to stay aboard.[2]

The lumber game in war time was played under new rules, government rules. For the first time mill owners looked on the federal government as its major customer and as a pervasive influence on its operations. Uncle Sam made the big decisions.

Lumber for cantonments was one of the first great demands. As the army and navy mustered in more than four million men, demand increased not only for barracks but for munitions depots, warehouses, hospitals, and housing for workers in the new war industries. More than a billion feet of lumber was purchased by the government within six months of the declaration of war; between five and six billion feet during the eighteen-month war. Since most military bases were in the South, most of the orders were for southern pine, but St. Paul & Tacoma supplied fir for Camp Lewis outside Tacoma and for a military base near Des Moines, Iowa.[3]

The Army Appropriation Act of August 29, 1916, known as the Federal Possession and Control Act, gave the president the power to take possession of any system of transportation in wartime should he consider it necessary. When an advisory body known as the Railroads' War Board (composed of five railroad presidents) failed to coordinate successfully the operations of the nearly seven hundred railroad lines in the country, President Wilson declared an emergency and, on the day after Christmas, 1917, took over the railroads.

A National Railroad Administration was created under Secretary of the Treasury William G. McAdoo, who took on the added title of director general of railroads. The Administration ordered a greater uniformity in equipment and standardization in repair work. One hundred thousand new freight cars were ordered. St. Paul & Tacoma, which had a thirty-year history of supplying boxcar material, received substantial orders. The new cars also helped relieve the acute shortage that had prevented shipment of lumber eastward.[4]

A tie shortage developed. John Foley, forester for the Pennsylvania Railroad, was put in charge of procurement.

As a young man Foley had worked in St. Paul & Tacoma camps. He favored Douglas fir as cross-tie material and ordered millions of feet of ties from Washington and Oregon mills, including St. Paul & Tacoma. For years after the war, Douglas fir dominated the tie market.[5]

In September of 1916 Congress had created the United States Shipping Board with power to regulate merchant shipping under the American flag and to promote the creation of shipyards and the construction of ships. The Board retained regulatory power and overall supervision of shipyard construction and shipbuilding, but active promotion was put in the hands of a subsidiary, the Emergency Fleet Corporation.

The E.F.C. was activated as soon as the United States entered the war. George Washington Goethels, builder of the Panama Canal, served as manager. Told that he must build 500 ships, Goethels said there was not enough steel. Told that some of the ships could be made of wood, he replied, "Birds are still nesting in the trees of which wooden ships are to be built." Lumbermen were up to that challenge.

The government pledged almost unlimited funds not only for the purchase of ships but to provide the capital required to create shipyards: $50 million as a starter, which was soon increased to $750 million and ultimately to 2 billion, 884 million. Four shipyards were built in Tacoma and twelve in Seattle for the manufacture of wooden freighters and transports.[6]

The war was only weeks old when St. Paul & Tacoma

joined with other Tacoma companies and capitalists to form the Tacoma Shipbuilding Company, which put its main shipyard next to the lumbermill. Tacoma Shipbuilding received government contracts that absorbed nearly the entire St. Paul & Tacoma output suitable for shipbuilding. Demand soon exceeded the mill's capacity.[7]

In October 1917, John Hewitt proposed that St. Paul & Tacoma enter an arrangement with a Tacoma group headed by John S. Baker under which the Baker interests bought the idle Ohop Mill, which had been designed to handle long timbers. They paid for having it dismantled and moved to Tacoma where it was set up on the foundation of the burned out Mill A. St. Paul & Tacoma agreed to supply the logs and run the mill for the next two years, paying the Baker interests a dollar per thousand board feet on all lumber cut. Shortly before the war ended, St. Paul & Tacoma bought the transplanted mill for $35,000.[8]

The government needed lumber for planes as well as ships. Aircraft frames were still made largely of wood. Mahogany was the preferred material for propellers, but there was disagreement on the merits of spruce, fir, and pine for the struts and beams. The Italians favored fir and more than 25 million feet of Puget Sound fir went into planes used in the Italian 1917 campaign against Austria. American engineers leaned toward spruce, which was lighter—it saved about seventy-five pounds in a training plane—and resisted splintering. But there were heavy problems in producing the lighter lumber.

The few concentrations of spruce were in remote areas, mainly on the Olympic peninsula. The remaining spruce was scattered through forests dominated by fir, cedar, and hemlock, which would have to be logged along with the spruce but for which there was less demand. Only about 15 percent of the spruce could be used for plane manufacture, and there was not much of a market for the lesser grades.

The greatest problem was labor. No one claimed that work in the lumber mills and logging camps was not rough and sometimes dangerous. In the fractured industry of logging and milling, with hundreds of owners, conditions varied from camp to camp and mill to mill. St. Paul & Tacoma along with Simpson and Weyerhaeuser were considered among the better companies to work for in the Northwest, but even in the best camps the hours were long, the pay low, even by the industrial standards of the day, and the authority of the boss often oppressive.

Though some men developed a love for the work, most of those in the woods were there because they could find no other jobs. They were men regarded as stronger of back than of mind, the unskilled, the foreigners, itinerants, "womanless, homeless, voteless." In times of low employment they had been almost powerless. During the war, able-bodied men were siphoned off into the military services and to higher paying jobs. In the spring of 1917 the men in the woods and mills felt themselves in position to make demands.

Unions were being organized by men of varied philosophy. The radical Industrial Workers of the World, known as the I.W.W. or the Wobblies, concentrated on recruiting the unskilled and the itinerant. They organized for immediate ends, but they talked of the ultimate overthrow of capitalism. The American Federation of Labor chartered two unions in the forest products field: the Shingle Weavers and the Timber Workers. Each was less concerned with demolishing the old and creating a new society than with getting a larger slice of the available pie.

The I.W.W. claimed about 3,000 members in Washington, most of them in the pine camps east of the mountains. The Shingle Weavers and Timber Workers unions claimed a combined total of 2,500 members, mostly on Puget Sound and the lower Columbia. In the spring of 1917 all three unions asked for a shorter day with higher pay and better working conditions. Though their demands were similar, their rhetoric was not. The Timber Workers asked a minimum of $3 for an eight-hour day in the mills, $3.50 for nine hours in the woods, and for better conditions, union recognition, the closed shop, and greater freedom from employer control at camps in matters not directly related to the job. (Some companies censored personal mail and forbade conversation in the mess halls.) The I.W.W. demanded the same work and pay gains as the Timber Workers did but talked of class war, of sabotage, of overthrowing the system. Such rhetoric inflamed both sides.[9]

The employers refused to negotiate. Loggers and mill owners, St. Paul & Tacoma representatives among them, met in Seattle and formed the Lumbermen's Protective Association. It went on record as resolving that the eight-hour day would hinder the war effort. The owners pledged themselves to maintain the ten-hour day. They raised a strike fund of $500,000 and threatened to fine any member operating less than a ten-hour shift $500 a day.[10]

The unions went on strike in July, shutting down four out of five mills and camps. For the first time pickets marched in front of the St. Paul & Tacoma mill. The company kept running behind the picket lines, though production slowed. The strikers stayed out until September, then under a barrage of criticism for "helping the Hun" they returned to the job, their demands unmet, but with a new tactic. The I.W.W. had devised the slow-down, or strike-on-the-job. "Conscientious withdrawal of efficiency" proved effective. By October, production of spruce was running so much below the requirements of the aircraft industry that the government intervened.

Just before the strike, Major Charles R. Sligh of the Army Signal Corps had been assigned to oversee spruce production. He named George S. Long of Weyerhaeuser to chair an Emergency Spruce Committee, which did little more than bemoan the situation. On September 6, the government commandeered all available milled spruce and diverted it to plane manufacture. And Major Sligh dispatched Colonel Brice P. Disque to the Northwest to report on the situation.[11]

There is still debate about Disque, a thirty-eight-year-old

*St. Paul & Tacoma's horse teams are mustered for a parade in 1910
(Cecil Cavanaugh Coll., Washington State Historical Society)*

West Point graduate who had dropped out of the army to serve as a prison warden in Michigan, but reclaimed his commission at the start of the war. Sligh later described Disque as a blow-hard and a disaster, some employers looked on him as the camel's nose of unionism, some unionists considered him a company tool. Whatever else, he was an innovator, a compromiser, an activist, and a very ambitious man.

Colonel Disque proposed, or agreed to, the formation of a kind of government-sponsored company union to be known as the Loyal Legion of Loggers and Lumbermen. Membership would include employers and employees, with all parties pledged "to support this country against enemies, both foreign and domestic." The domestic enemy, in Disque's mind, was the I.W.W. The 4-L, as the group styled itself, was not to be a labor union "in the common acceptance of the term" but a "patriotic unit," which would neither organize workers into unions nor disrupt "any legitimate labor unions." Definitions were imprecise, there was a lot of room for interpretation, and Disque did the interpreting.[12]

In theory, power in the 4-L was balanced equally between workers and management. They had conflicting interests. Somebody had to decide. Usually it was Disque. At the start he relied heavily on the advice of Henry Suzzallo, the president of the University of Washington and state defense chairman, and Carleton H. Parker, a professor of economics at the university. Each made the point that if spruce was to be logged and milled, many of the workers' demands would have to be met.[13] The great issue was the eight-hour day. Disque imposed it, and followed that up with orders to employers to furnish bunk-houses with clean bedding and a weekly change of sheets and pillow slips at a charge of a dollar a week.[14]

Still there were not enough men working in the woods. Disque's answer was to create the Spruce Production Division, one of the oddest outfits ever activated by the United States Army. It was a body of soldiers assigned to cutting spruce and bringing the logs to the mills. They received the going civilian wage for their work, minus their Army pay and minus $7.59 a week as ration allowance. Some 7,000 soldiers volunteered for the Spruce Division and another 23,000 were assigned. In theory, the division took only experienced loggers; in practice quite a few of the wooden soldiers couldn't tell a Swedish fiddle from a deacon seat. Some never before had raised a callus but they worked very hard—the division produced 143 million feet of spruce to the saw—and they were not without casualties. The woods are dangerous, especially for amateur loggers, and during the flu epidemic of 1918 the men in the rain forest were at great risk.[15]

During their first winter in the woods, most of the soldiers were assigned to camps near existing roads and set to cutting down spruce and splitting them on the spot. This was as inefficient a way of getting lumber as could be devised, short of sabotage. Major Griggs, who had taken leave as president of St. Paul & Tacoma to serve in the Coast Ar-

By the end of World War I, horses had been replaced by gasoline combustion engines for hauling lumber. A truck pulls three wagons of slabs up an 8 percent grade on Delin Street to the furnaces of a steam laundry (Cecil Cavanaugh Coll., Washington State Historical Society)

tillery, was transferred to the Spruce Division.[16] He helped correct a number of inefficiencies and recommended construction of the Spruce Railroad, a line designed to service existing operations and a proposed new mill. (Nineteen days after the war ended, it was ready to carry logs.)

After only a few months with the Spruce Division, Griggs was named to succeed J. H. Bloedel in charge of a program to find ways of using fir in airplane construction. The effort paid off. "Fir is now entering upon its own," Griggs told a meeting of West Coast lumbermen in September. "It is equal to spruce for aircraft material and in some respects superior to it and is being so recognized." By the end of the war, 40 percent of the lumber being cut for planes was fir.[17]

Looking like buildings, stacks of boards wait beside the tracks for shipment (from West Coast Lumberman*)*

State of the Art: Mill C

WITH THE Armistice, the building of wooden ships came to a sudden halt. Airplane construction continued but the government had little interest. The American Expeditionary Force was brought home and disbanded; its cantonments were taken down or boarded up. St. Paul & Tacoma turned again toward civilian production in a changed world.

A revised board of directors guided the company. Addison Foster had died just before the war.[1] Harrison Foster (he had adopted the name Harrison because his wife thought the baptismal Harry too common) sold his stock to the Griggs family in 1919 and moved to Kentucky. George Browne had sold his shares and returned to New York.

Henry Hewitt died on May 2, 1919, at the age of seventy-seven, following surgery. At the time of his death he was treasurer of St. Paul & Tacoma; president of Wilkeson Coal and Coke; president of the Bank of Oregon at Coos Bay; president of the Hewitt Land Company, which owned more than 2 billion feet of timber in Washington, British Columbia, Idaho, Oregon, California, Arizona, Arkansas, Missouri, Kansas, Minnesota, and Wisconsin; president of the Hewitt Investment Company, which also owned much timber and mineral land in Oregon; and president of the Sumpter Lumber Company, with 30,000 acres of timber in the Northwest. It must have pleased him, staring at the white ceiling of his room in Tacoma General Hospital, to think he probably owned more timber than any individual in the world, having outlasted the late Czar Nicholas of Russia. Perhaps he remembered that first payment in white pine for work done on the Fox and Wisconsin Improvement.[2]

On the post-war board of St. Paul & Tacoma, the Griggs interests dominated. Major Everett Griggs was president, his older brother Herbert was secretary, Leonard Howarth was both vice-president and treasurer. Chauncey Milton Griggs of Minnesota was on the board, as were C. H. Jones, the last of the originals, and Will Hewitt.

The basic assets of the company remained the forest realm purchased by the founders from the Northern Pacific, and the acres of technology assembled on the tidal land they had reclaimed for industry. Much of the original stumpage had been logged but 55,000 acres, more than half, remained uncut. To this had been added, largely on the advice of Hewitt, 15,000 acres by purchase or trade. The 1920 holdings amounted to some 70,000 acres, compactly blocked in an area of 150 square miles in southeastern Pierce County.

Timber cruisers estimated that St. Paul & Tacoma had more than 2 billion feet of standing timber, 92 percent of it Douglas fir, the rest, cedar and western hemlock with a scattering of spruce. Hemlock, long scorned as a weed because it bore the same name as the pitch-heavy eastern variety, had yet to establish its value; the cedar was superb. Company foresters estimated that even if St. Paul & Tacoma cut only its own timber instead of buying a third of what they sawed as was company practice, they could supply customers with 100 million board feet a year for the next quarter century. If judiciously harvested, the cut could go on "indefinitely."

The edge of the virgin forest lay twenty-five miles from the mill. The section of the Tacoma, Orting and Southeastern Railroad between Kapowsin and Orting had been transferred to the Northern Pacific, but the company owned the eight miles of track it had pushed into the foothills beyond Kapowsin. The rails gradually rose from switchback to switchback to climax in an interchange at Camp Four. From there spur lines extended to logging operations and to Camps One and Three, higher in the foothills, each at an elevation of about 1,600 feet. In all there were about twenty-five miles of logging railroad. Between seventy-five and eighty flatcars, each bearing from 8,000 to 10,000 board feet of logs, rolled through Kapowsin each day on their way to Tacoma.

Master of this feast for the saws was Will Hewitt, Henry's oldest son, who was raised with white pine and came of age amid Douglas fir. Assisting Hewitt as general superintendent of logging was Andrew LeDoux. Will and Andy had worked out a system.

Before operations began in any area, it was cruised thoroughly. Charles A. (Chuck) Billings had charge. Each forty acres was divided into 2½-acre tracts, their locations marked by blazes on trees. The altitude of the area was established from bench marks set earlier by government surveys, or those of other engineers.

Knowing where he was both horizontally and vertically, the timber cruiser checked his aneroid barometer to determine variations in elevation as he surveyed a tract. Trees were counted and noted as to species and scale. Their condition was recorded. The finished estimate included a list of fir, cedar, spruce, and hemlock for each subdivision, with size, age, length, and soundness judged. The resulting map, showing elevations as well as estimating timber, helped the locating engineer and head logger plan their moves through the forest.

Camp Four was the construction camp. Crews from the pile driver, steam shovel, and gravel pit, and railroad construction teams spent their nights there. Camp Four, or the mobile Camp Two, which was built into railroad cars, was advance headquarters as well for surveyors; there were work tables for laying down what had been found, a phone system to control the movement of logging trains and speeders, and an office for the timekeeper—that hated Cyclops. The company store, mounted on wheels, moved out from Four on monthly schedule, stocked with longjohns, snoose, witch hazel, and other amenities.

Buildings in the camps were mounted on log skids. They could be towed into place, or picked up by logging cranes and deposited on flatcars for movement along the line. Home was where you found it.

Eight men shared a bunkhouse. Each had his own iron bed. The company supplied blankets, sheets, and a flunkey who was supposed to swamp out and make up the room. A big iron stove dominated the center of each bunkhouse. There was never a shortage of wood to burn. Water was run from a central tank to each building, and electric lights had replaced the kerosene flickerers. Each camp had its blacksmith shop, oil house, warehouse, and filing room where saws were kept in file for the fallers and buckers. Dropping Douglas fir was still a matter of muscle and filed steel; the chain saw was two decades in the future.

St. Paul & Tacoma had adopted high-lead logging, an extension of a process devised by Horace Butters of Maine for use in the white pine country of Michigan. A huge apparatus, which the inventor modestly called the Horace Butters' Patent Skidding and Loading Machine, used a wire and trolley strung between two tall trees, to raise logs off the ground as they were being moved. Butters' innovation

W. H. Hewitt (left) and Andrew LeDoux, superintendent in charge of logging, look over average old-growth fir logs (from American Lumberman, May 21, 1921)

By 1921, three St. Paul & Tacoma camps were lighted by electricity. All of the buildings shown here could be loaded on flatcars and moved down the line to new logging sites (from American Lumberman, May 21, 1921)

did not gain popularity in the white pine country, but southern loggers found it useful for hauling cypress out of swamps. It was in the western forests of fir and redwood that high-lead logging found its home.

C. H. Jones used a variant of the Butters' system in his Grays Harbor operation, and his colleagues in Pierce County took up the idea. The equipment was Bunyanesque: pulley blocks weighed a ton, spar trees topped off at 200 feet, the logs were 40 feet long and up to 8 feet through. A visitor from the East described the high-wire act:

A 1½-inch steel cable—the sky line—about 1500 feet long is suspended between the tops of two tall trees known as spar trees, which are held rigid by guy lines. Upon this sky line is operated a great trolley. By means of a rehaul system of lines the main line and trolley are carried out to where the logs are to be picked up and brought to the railroad for loading. Down from the trolley the skidding line leads to the ground and also leads away from either side of the sky line a distance of about 200 feet. The skidding line is attached to the chokers, which pass around the logs to be brought in. The hoisting engine lifts one end of the log high in the air and it is dragged to within the loading distance of the railroad, where it is picked up by the auxiliary loaders and put directly on the waiting cars. These yarding engines will load on the cars an average of from 80,000 to 100,000 feet of logs a day.

The mills, too, were being modernized. Early in 1919, Major Griggs hired Earl M. Rogers as assistant general manager and assigned him the task of renovating Mill B and Mill C.

A high-lead yarder with a donkey engine in foreground and Lake Kapowsin and the Tacoma and Eastern Railroad track along the far shore (from American Lumberman, *May 21, 1921)*

Rogers was an old pro at putting up sawmills. A native of Wisconsin and a millwright since boyhood, he had come west to work for Weyerhaeuser in Idaho. Afterwards he superintended construction of the big Weyerhaeuser mill at Everett. It was Rogers who, on April 29, 1915, checked the line of incoming logs and gave the signal to George S. Long to turn on the current, which marked the start of the Weyerhaeuser company cutting its own timber in its own plant. Rogers was in charge of Mill B at Everett until June of 1918 when he had been hired to build a mill at Port Angeles for the Siems—Carey—Kerbaugh Corporation as part of the spruce production program. There he faced a labor shortage, floods, and an influenza epidemic of such proportions that the commander of the Spruce Division wired his headquarters to "either give orders to evacuate or forward a trainload of coffins." Even so, Rogers, in five months, had the mill nearly ready to start cutting—only to have peace break out. The contract was cancelled, the mill abandoned.

Rogers moved on to Seattle where he helped form the Rogers—Mylroie Lumber Company, which bought a small mill on Lake Union. In April of 1919 St. Paul & Tacoma lured him from Seattle with the promise of a chance for more mill construction. He stayed with the company the rest of his life.

In a high-lead logging show, the logs floated through the air, but with no great ease. A high-climber first limbed and topped a spar tree to serve as focal point for the cables (Kenneth Brown, photographer; author's collection)

First Rogers rebuilt Mill B. His aim was to increase production, cut operating costs, and improve the product without interrupting the two shifts on the job. It was done in twenty months. The key change was from steam power to electric motor drive throughout the plant (the Major's proposal of 1913), a transformation that increased flexibility in the entire operation.

St. Paul & Tacoma generated its own power from a plant that Major Griggs had earlier built adjacent to Mill C in the center of the tideflat property. Waste from the mills fueled the boilers. Sawdust and slash were carried to the storage house by belt conveyor from Mill B and by chain conveyor from Mill C and by blower from the planing mill. Another conveyor system carried the fuel from the storage house to the boiler plant, which held a dozen Stirling boilers capable of generating more than 5,000 horsepower.

The generating plant, with a 2,500-kilowatt turbine and a 1,000-kilowatt turbine, stood alongside the boiler plant. The turbines were connected to a 3,750-kilowatt step-up transformer that boosted voltage to 4,000 for outside transmission to various departments, where it was cut down to motor voltage of 440. Direct current for operating cranes, electric motors, and battery charges was furnished by a 75-kilowatt turbine and a 100-kilowatt motor generator.

With this equipment St. Paul & Tacoma not only met its own power needs but supplied electricity and steam to several other industries on the tideflats, including Carstens Packing Company, the Wood Handle Company, and the Puget Sound Iron and Steel Works. The electric plant was tied in with Tacoma City Light for interchange of current. When water was low behind the city's dams on the Nisqually River, St. Paul & Tacoma could loan the city electricity.

Additional waste from the mill was sold to the Consumers' Central Heating Company, which for sixty years furnished steam heat to downtown Tacoma. St. Paul & Tacoma had conceived the idea of warming up Tacoma business establishments and offices with its mill waste. The company owned stock in Consumers' Central, as did some of its officers individually. A belt conveyor, 630 feet long, carried hogged fuel from Mill B over a city street and some railroad tracks to scows on the Puyallup. From there it moved to the heating plant on the waterfront. The conveyor was unusual in that, except for the two-foot-wide belt and the roller bearings, it was entirely of wood.

Little waste was wasted. Slabs were sorted and directed to the lath mill part of Mill B, where a five-saw bolter and a four-saw stripper and trimsaw turned out 75,000 feet of laths in an eight-hour shift. One man could handle the entire operation. Sawdust and shavings went to the steam plants. Other trim was carried to the fuel department, which sold wood to the public, delivering the loads first by horse-drawn carts, later by truck.

Though the little things were not overlooked, the point of Rogers' rebuilding of mills B and C was to facilitate the flow of logs and lumber through the plant. The ponds at Mill B and Mill C were enlarged and deepened, as was the channel connecting them with the storage areas on the bay.

Waste from the sawmill operations was burned to generate power for mill operations. The surplus steam and electricity was sold to other industries on the tideflats (from American Lumberman, *May 21, 1921)*

Logs brought down from the mountains were dumped into the ponds, those destined for specialty cuts going to Mill C, the rest to Mill B, where they faced either the 10-foot band mill or the 9-foot double-cutting band mill, each driven by a 250-horsepower electric motor.

As the lumber emerged from its first encounter with steel it was carried on feed rollers through one of the most automated manufacturing lines in the industry. A combination edger, spotter, and transfer took care of the cants. A 44-foot-span steel rollercase handled lumber coming off the edger and discharged it either onto flair chains that carried it to the trimmer, or to a hydraulically operated rollercase that took it to the rear of the mill. There the timbers were picked up by a three-ton electric crane for loading into industrial cars—small freight cars pulled by electric locomotives over the twenty-seven miles of track in the St. Paul & Tacoma tideflat area, or they were put onto gravity rolls that carried them to the timber deck.

In the remanufacturing plant adjoining the trimmer a battery of saws (60-inch horizontal double resaw, 84-inch resaw, 24-foot undercut trimmer, 6-inch pony edger, and 24-foot slasher) awaited the decision of the sawyer about the cuts to be made. Three sorting tables arranged parallel to each other furnished nearly 1,000 feet of sorting space on which the pieces were arranged by size and grade.

Lumber coming off the sorters was piled into units to be picked up by grasp-carriers that moved along an overhead monorail suspended from fifteen-inch I-beams that linked the major plants on the campus. Other stacks were loaded into industrial cars and hauled to one of the seven dry kilns. The lumber was automatically stacked on edge by an electric powered machine as the car on which it was riding rolled into the kiln. As it came out the far end after the drying process the car took the treated lumber to the unloading machines at the rough dry shed, where it was placed on a sorting chain, re-graded, and sorted into unit packages that were picked up by the monorail and taken to a storage shed or to the planing mill for processing by the woodworking machines.

The planing mill was the one Major Griggs had built to replace that destroyed in the great fire of 1912. Covering an area as big as a city block, it could handle 400,000 board feet a day. Equipment included a paving block machine; ten cut-off machines; four 15-inch flooring machines; a 9-inch flooring machine; two four-side 6-by-24-inch surfacers; a four-side 20-by-24-inch surfacer; a combination re-saw and surfacer; two molding machines, a rip saw and roller re-saw. Specialties of the plant were Douglas fir flooring and cedar siding.

After planing, the lumber was trucked to the shipping shed between the planing mill and the thirty-five-acre storage yard beside the railroad tracks. Dressed lumber was put into railroad cars immediately or stored in the dressed lumber sheds.

Shed 7 was chiefly a loading shed and little lumber was stored there. Railroad tracks ran through it so lumber could

A 22-saw pneumatic trimmer in Mill B made the final cuts as the lumber passed down the line
(*from* American Luberman, *May 21, 1921*)

213   *State of the Art: Mill C*

A worker in the storage shed stands beside an unblemished board, 4 feet by 24 feet
(*from* American Lumberman, *May 21, 1921*)

be put directly onto the cars, fourteen of which could be under the roof at the same time. Shed 8 held an automatic stacker. It usually contained from 200,000 to 300,000 feet of dressed lumber. Shed 9—144 feet wide and 600 feet long—was used to store clear lumber after planing. The tracks ran along one side of this shed, the roof extending over the cars. It could hold eight million board feet of dry lumber. Shed 10 was bigger, with a capacity of ten million feet. It had room as well for the biggest of all sorting tables on which lumber was put into unit packages according to size and grade.

The shingle mill, standing alongside Mill B, could cut 200,000 shingles in an eight-hour shift. Cedar logs were snaked from Pond B up a slip to the deck where a swing cut-off saw and a steam splitter worked them into bolts. A chain conveyor carried the bolts into the mill proper where shingle-weavers facing the blades of six upright machines cut them into roofing material. Two concrete kilns, 20-by-100 feet, dried the shingles, after which they were moved to a storage shed which could hold nine million.

Lumber that did not go through the kilns was stacked for air drying in the main storage yard. Double tracks ran through a pattern of twenty-one alleys criss-crossing the yard. They allowed the industrial cars easy access to areas where various cuts and grades were concentrated.

St. Paul & Tacoma had been conceived as a rail mill and most of its product now moved away on tracks. All rail lines serving the Pacific Northwest were connected with the plant, including the Northern Pacific, the Great Northern, the Burlington, the Oregon–Washington Railroad and Navigation Company (a part of the Union Pacific system), and the Chicago, Milwaukee and St. Paul. Though the company had special contractual obligations with the Northern Pacific because of the original timber purchase, it worked closely with the other main lines. When the Union Pacific extended service to Tacoma, the St. Paul & Tacoma Lumber Company provided it with thirty acres of terminal space—and gained access to eleven thousand miles of new territory for lumber sales. Chicago, Milwaukee and St. Paul built its own bridge across the Puyallup to connect with the company's own rail tracks.

The cargo trade—shipment by water—remained important. Mill C, which had been set up originally to supply long lumber for shipbuilding, now specialized in heavy timbers, many of which were too long to be moved by rail and were sent by ship. Much of the product of the creosoting department, which had been established in 1912 at Hewitt's urging, also crossed the wharf. The original 130-foot retort had attracted national notice in treating the pilings and stringers for the high bridge across City Waterway at Eleventh Street. Orders were still coming in for treated lumber to be used in salt-water construction. Creosoted wood blocks for street paving remained an item of export, especially to areas around the Pacific rim where the auto had yet to replace the horse.

The cargo shipping dock, 800 feet long and 250 feet

Bridge stringers move into the creosoting plant for preservative treatment. Its 130-foot-long retort could process 40,000 board feet a day (from American Lumberman, *May 21, 1921)*

wide, fronted on Commencement Bay. Water alongside the dock was 35 feet deep at low tide. Three ocean-going steamers could take on cargo simultaneously. No dock on the Pacific Coast offered faster turn-around time. Two standard-gauge railroad tracks, two 36-inch gauge industrial tracks and a 20-foot crane track ran the length of the dock. A 76-foot electric whirly crane capable of lifting five tons at its extreme reach and much heavier loads at shorter distance could operate at any point along the dock.[3]

Across the bay, around the point, lay a world hungry for lumber, and beyond the mountains to the east, a nation.

The creosoting plant with an array of treated piling in the foreground. The plant supplied some of the material for Tacoma's Eleventh Street bridge, which was originally paved with creosoted wood blocks (from American Lumberman, *May 21, 1921)*

Tokyo, New York, Sunnyside, and Way Points

WITH THE rebuilding of Mill C, St. Paul & Tacoma had the capacity to produce nearly a million board feet during a double-shift day, not counting laths and shingles. The problem was that between 1921 and 1925 the production of lumber in Washington more than doubled as big mills retooled and small operators returned to cutting. Competition for the market was intense.

St. Paul & Tacoma's export trade, like that of most West Coast lumber companies, was handled largely through the Douglas Fir Exploitation and Export Company, which had recovered from its wartime slump. In April of 1918 Congress passed the Webb–Pomerene Act, exempting from the restrictions of the anti-trust laws any U.S. associations engaged solely in export trade. The DFE&E was reorganized, its headquarters were moved from San Francisco to Seattle, and a major marketing campaign was launched.[1]

Japan, at that time a nation of eighty million people for whom wood was the dominant construction material, offered an especially attractive market. Japanese buyers preferred shipments that could be resawed. They ordered even-sized lumber ranging from 4-by-4 to 24-by-24 inches. Timbers 30-by-30 were purchased when available.

To the men at St. Paul & Tacoma, and other mills filling these orders, the smaller cuts came to be called "baby squares" and the larger ones "Jap squares." The "Jap squares" were the biggest pieces of lumber that could be taken out of logs. The sawyer simply cut off slabs to square up a log of full length, usually forty feet. The DFE&E promoted them heavily, describing the squares as "sound, strong lumber free from shakes and large knots, loose knots or rotten knots, and defects that materially impair strength, well manufactured and suitable for good, substantial construction purposes, also suitable for remanufacturing into smaller sizes without unreasonable loss from defects."

DFE&E accepted orders at a 2½ percent commission. Payments for export orders were usually arranged by letters of credit between the buyers and the export company. Banks would honor the letter upon presentation of the mills' invoices accompanied by certificates from the Pacific Lumber Inspection Bureau as to grades and volume. The DFE&E could deal with foreign companies more effectively than individual operators, who could be played off against each other and tempted into rate cutting. And since DFE&E chartered the bottoms necessary for its shipments, St. Paul & Tacoma did not have to concern itself with finding vessels to deliver lumber to foreign ports from its Tacoma docks. By 1923 the agency was contracting for 40 percent of Japan's lumber imports, with St. Paul & Tacoma and Long–Bell of Longview getting lions' shares among the 109 members.[2]

On September 1, 1923, one of history's most disastrous earthquakes shook down the heart of Tokyo, leveled Yokohama, started widespread fires in other communities, claimed an estimated 100,000 lives, and left millions homeless. Four hundred thousand buildings had to be replaced. Japan immediately asked shipment of 18 million feet of boards, planking and flooring; 66 million feet of baby squares and flitches; 60 million feet of Jap squares; and 120 million board feet of logs and piling. The very streets of Tokyo had burned. St. Paul & Tacoma received a request for "all the creosoted blocks of Douglas fir paving you can supply." The Japanese embassy in Washington, D.C., closed negotiations with the DFE&E for 96,400,000 feet of Douglas fir and hemlock.

The Japanese government created a Metropolitan Construction Board to supervise rebuilding. The Board asked Secretary of Commerce Herbert Hoover for assurance that there would not be a repetition of the price rises in lumber that followed the 1906 quake in San Francisco. The West Coast Lumbermen's Association circularized its members urging restraint. Prices did surge briefly to their highest point since 1920, average $27.26 per thousand at the mills, but quickly dropped to only 10 percent above pre-quake levels.

Major Griggs visited Japan early in 1924 on an inspection trip. He estimated that it would require two billion feet to rebuild Tokyo alone but warned that this did not necessarily mean prosperity for West Coast mills, and he proved to be correct.[3]

Japanese harbors were disaster areas. Ships lay out in the stream for weeks, waiting opportunity to unload. When lumber was put down ashore, it could not easily be moved. Rail lines had been wrenched apart, canals blocked, barges wrecked. It cost ten dollars a thousand to get lumber from Tacoma to Yokohama, seventeen dollars a thousand to move it from Yokohama to Tokyo. With Japan's own industries crippled, the vast imports dislocated the balance of trade. The yen plunged against the dollar. Banks refused to issue letters of credit. The West Coast Lumberman reported that orders had plunged "almost to the zero point," earlier shipments remained in the hands of the importers, and "just what has happened will never be completely known."[4]

Eventually Japan returned to its status as a major customer. (Major Griggs remarked in a letter to fellow lumberman Mark Reed in 1927 that "every operator in the country is interested in the development of the lumber market in Japan.")[5] But the industry had been through one more cy-

"Jap squares" were trimmed to be cut to Japanese standards in Japanese mills
(Marvin D. Bolland, photographer; Tacoma Public Library)

cle of high demand, quick expansion, and almost instant over-production. The quake boom which dominated all lumbermen's thoughts for a year ended almost as quickly as it began.

A similar but less dramatic situation followed the opening of the Panama Canal to regular civilian shipping in 1920. Anti-trust laws precluded the operation of an industry-wide agency similar to the DFE&E to distribute lumber on the East Coast. Producers adopted varied strategies in their hunt for trade beyond the Canal. Some companies sold most of their lumber to firms such as Dant and Russell of Portland, who controlled large amounts of steamship space. Others sold to local wholesalers who in turn sold to their eastern counterparts. A few large companies such as Weyerhaeuser and Long–Bell maintained their own yards at Atlantic Coast cities. Still others preferred to station agents in key East Coast centers to solicit wholesale orders.[6]

St. Paul & Tacoma used a combination of approaches. The first of the company's lumber to move through the Canal went on ships of the Calmer Line, the American–Hawaiian Line, the Matson Line, and the Luckenbach Line, which had carried iron and steel to the West Coast and came to Tacoma looking for return cargoes.

When difficulties arose in finding adequate wharfage and storage in the East, St. Paul & Tacoma entered into association with the A. C. Dutton Lumber Corporation, which had available space. Shipments could move directly from the St. Paul & Tacoma dock to Dutton's deep water dock at Poughkeepsie on the Hudson. Later a second distribution yard was established at Providence, Rhode Island.

To boost sales in the East Coast market, St. Paul & Tacoma joined several lumber companies and steamship lines in organizing the Pacific–Atlantic Lumber Corporation—Palco, for short. Sales headquarters were in New York, under the management of George Corydon Wagner, Jr., the son of Dr. George Wagner and Colonel Griggs's daughter, Heartie.

Born in Tacoma in 1895, Cordy Wagner was graduated from Yale just in time to volunteer for the First World War. On his return from overseas in 1919 he went to work for St. Paul & Tacoma in the traditional starting block for family members, as timekeeper. His interest lay more in finance and sales than in operations, and he was channeled into the business side of management. At Palco, his first important assignment, he could call on the advice of Lynde Palmer, who had represented St. Paul & Tacoma and Wheeler, Osgood and Company in New York since 1900. Wagner credited Palmer with steering him toward his first big success, the sale of a huge shipment of stringers for use in New York subway construction. Subways became a considerable outlet for Douglas fir.

While the bulk of Palco orders were from distribution yards in the metropolitan areas of such major seaboard cities as Boston, Philadelphia, and New York, Wagner developed a considerable volume of business by arranging for backhaul shipment by rail to Ohio, Indiana New York, and Pennsylvania. In this he was aided by the company's existing network of sales representation.* [7]

An aggressive sales organization in the Midwest was especially important at this stage because of prolonged litigation and negotiation between the Northern Pacific and St. Paul & Tacoma over interpretations of the original contract.

One provision was that the carrier agreed to furnish log cars, delivered to the interchanges in the woods, and to haul the loaded cars from the interchanges to the mill ponds in Tacoma at one dollar a thousand feet, log scale. St. Paul & Tacoma was to take delivery of the empties at the interchange, haul the cars to the logging areas over its own trackage, load them, and deliver them back to the interchange. The Northern Pacific cancelled this provision unilaterally and began charging the rates set by the Washington Railroad Commission for intrastate log shipments. St. Paul & Tacoma challenged this action in court, suing for reinstatement of the contract rate. The railroad prevailed.

The contract also required St. Paul & Tacoma to ship over the NP lines a percentage of lumber manufactured from timber taken from the NP land grant. St. Paul & Tacoma had made additional purchases from Northern Pacific, as well as from other timberland owners after 1888, purchases that had greatly increased its inventory of timber and volume of production over what had been estimated in the original contract. The company and the carrier disagreed on the volume St. Paul & Tacoma was committed to ship over the NP lines each year. After prolonged negotiations, lasting years, and the threat of additional litigation by both sides, Herbert Griggs and the railroad attorneys reached a compromise agreement establishing the volume of timber remaining on land purchased from the Northern Pacific and setting the amounts that had to be shipped over NP tracks. [8]

The coastal trade, the mainstay of the company's early years, remained important. Shipments to California were handled through the Consolidated Lumber Company, which St. Paul & Tacoma had created in 1905 in combination with the W. H. Perry Lumber Company of Los Angeles and the Charles Nelson Company of San Francisco.

Consolidated maintained a large planing mill and a wholesale distribution yard in the harbor of San Pedro. The company yards adjoined docks for handling coastwise vessels and had connections with the Southern Pacific, the Los Angeles and Salt Lake, and the Pacific Electric Inter-urban, as well as with the network of highways which, in the 1920s, was becoming a feature of Southern California life.

*In the 1920s this included John McDonnell in Chicago; John H. Burnside, a one-time sales manager in Tacoma, as head of an independent company that handled the Colorado, Wyoming, and western Nebraska territory; G. M. Barber of Spokane, who covered Montana and part of Wyoming; H. I. Isbel in Indiana, southern Michigan, and northern Ohio; the J. D. Merrill Lumber Company of Sioux City in South Dakota and Iowa; Morrison, Merrill & Co. of Salt Lake in Utah and southern Idaho; Dant & Reynolds of Detroit, representing the company in southern Michigan; Earl Randall of Beloit in Wisconsin, northern Illinois, and northeastern Iowa; J. C. Summers Lumber Company of Omaha in most of Nebraska; the Wigate Lumber Company of Kansas City for Kansas; and Arthur Hawksett of Minneapolis in the Minnesota area.

Cargo dock showing two steamers being loaded at one time. The electric railway terminating here led to every shed, yard and mill on the St. Paul & Tacoma campus. The insert shows the latest thing in cargo cranes in 1921 (from American Lumberman, *May 21, 1921)*

East of the Cascades, there were the company yards. During the depression that followed the Panic of 1893, when many retailers went out of business, St. Paul & Tacoma had experimented with running its own retail outlets. The first opened at Yakima in 1894 and proved successful. Others were added as irrigation projects, private and federal, opened new areas for development. At the maximum, the company operated nineteen yards. These were eventually consolidated into fifteen.

The yard system was a drain on the managerial energies of company officials. The solution proved to be John Dower, a self-made businessman from Minnesota. Born on a farm in Michigan during the Civil War, Dower left school at the age of twelve (he was only in the fourth grade) and went to work on a farm in northern Minnesota. He drove oxen for four years, breaking sod in summer, hauling cord wood, ties, and saw logs in winter. It was winter work that led him into the retail lumber business and brought him into contact with Hewitt and Jones. Dower specialized in retailing lumber that others had manufactured. He formed the Dower Lumber Company at Wadena, Minnesota, on January 1, 1900.

Dower was so good at peddling boards that, in 1919, St. Paul & Tacoma joined him in forming the John Dower Lumber Company, of which he and St. Paul & Tacoma were equal owners. All St. Paul & Tacoma retail yards, except the one in Tacoma, were turned over to the John Dower Lumber Company. Eleven lay along the NP tracks at Selah,

Toppenish, Grandview, Yakima, Prosser, Kennewick, Moxee City, Naches, Wapato, Sunnyside, Outlook, and Richland. A yard at Finley was on the Spokane, Portland and Seattle line, and the ones at White Bluffs and Hanford (later to generate unexpected construction and destruction possibilities) on the Chicago, Milwaukee and St. Paul line. Dower moved to Tacoma in 1921 and managed the yards successfully until his death nearly a quarter of a century later.[9]

In Tacoma, the St. Paul & Tacoma retail yard stood on the campus directly west of the dressed lumber storage yard. A 48-by-150-foot shed covered finished lumber offered for retail sale. An overhead girder-type crane with a 38-foot reach swung boards from the little industrial cars, which were confined to the tideflat area, onto wagons or trucks for delivery.

The introduction of the overhead monorail and the narrow-gauge industrial cars had greatly reduced St. Paul & Tacoma's reliance on horsepower for movement of lumber around the plant. The old barn had been torn down during the expansion of storage area, and a garage was built alongside the machine shop with a capacity for fifteen trucks. A small garage was built later for storing the tin lizzies of the few employees who did not rely on public transit service, which was then frequent and cheap in Tacoma.

Overall management of sales rested with A. H. Landrum, a refugee from journalism. Landrum had been working for the *American Lumberman* in 1911 when its editor, Leonard

St. Paul & Tacoma's retail yard, showing the city in the background, ca. 1900 (Washington State Historical Society)

Bronson, left to become secretary of the National Lumber Manufacturers Association during Everett Griggs's presidency. Landrum came west with Bronson. By the time the NLMA offices returned to Chicago under the next president, Landrum was hooked on Puget Sound. He joined the St. Paul & Tacoma sales department in 1913 as assistant manager and became manager at the end of World War I.

Besides selling lumber, St. Paul & Tacoma operated a large wholesale and retail general store. It stood on St. Paul Avenue between the plant and the general offices. It offered hardward, groceries, men's wear, and mill, logging camp, and mine supplies. The operation reflected Colonel Griggs's love of the grocery business and his insistence that while volume was important, so was the convenience of the customer.

A side track from the railroad ran behind the store. Wareroom floors were built so their height was level with that of the boxcars. Three cars could be loaded and unloaded simultaneously. This made it possible for the store to send supplies to mills, camps, and mines on short notice at minimum expense. The store was also located only a few yards away from the timekeeper's office, so employees, on being paid, could stop to shop, a convenience that benefited both company and employee. The St. Paul & Tacoma company store was, for some years, the largest merchandizer in Tacoma—and the first to use automobiles to deliver purchases.

The company also maintained the St. Paul and Tacoma Hotel, in reality a boarding house sometimes called Hotel de Gink, for the convenience of its employees who did not own their homes. (With low land prices after 1893, cheap lumber, expansive city limits, and good streetcar service, Tacoma had become the leader among industrial cities in the U.S. in percentage of home ownership.) The company hostelry was a building of 80-by-100 feet, four stories tall. The first floor held a lobby, reading room, and billiard and pool room, as well as a dining room that proved suitable for entertaining such groups as the delegates to conventions of the National Association of Railroad Tie Producers and the American Wood Preservers Association. On the top floors were the men's rooms. Each floor had two baths, two toilets, and a room for drying clothes.[10]

Keeping Up with the Joneses

CHARLES HEBARD Jones, the last of the founders of St. Paul & Tacoma, died on November 28, 1922, at the age of seventy-seven, after a long illness.[1]

The taciturn Vermonter, a man who learned by doing, had brought great changes to the area he came to in his mid-forties. He introduced the band saw, with which he had become acquainted in Michigan and Wisconsin, to the Douglas fir mills of the Pacific Northwest. He demonstrated, too, that the level but soggy land at the mouth of the Puyallup River could be firmed up and made to support buildings containing heavy machinery, thus making possible today's industrial complex on the Tacoma tideflats. Yet Jones had never been fully committed to Tacoma.

While there remained white pine to be cut along the Menominee, he spent half of each year in Michigan, supervising the operations of the Ramsay and Jones Mill on the western shore of Lake Superior. After closing down the Michigan mill, he bought controlling interest in the historic North Western Lumber Company on Grays Harbor, which he operated until his death.

The North Western was the first steam-powered mill on the harbor. It had been built in 1882 as a joint effort by Captain A. M. Simpson, a San Francisco ships captain and lumber dealer, and G. H. Emerson, a millwright. Simpson donated the mill site ("the finest opportunity on the coast") and Emerson, operating almost without capital, scrounged up some machinery from a defunct operation in Mill Valley, California, brought it north on the brig *Orient,* located a pile driver, diked and diverted three languid creeks that oozed across the tideflats at the mouth of the Hoquiam River, and created the mill. For more than a decade its entire product went to San Francisco to be sold through the Simpson Lumber Company. The plant burned in 1896, but was immediately rebuilt with the addition of a planing mill, box factory, sash and door cutting plant, and dry kilns.[2]

Jones's interest in the Grays Harbor area had been aroused when he and Hewitt visited there in 1887 before their meeting with Griggs and Foster in Tacoma. He and Hewitt began buying timberlands in the early 1890s, and controlled some stands of exceptionally large trees. After Jones took charge, the mill specialized in the shipment of large and long timbers to the East Coast, one shipment being a single timber forty inches square and eighty feet long, loaded on three flat cars. The mercantile department of the mill became the largest of its kind on Grays Harbor, much of the foodstuff it distributed coming from Pacific Meat (later Carstens) and the West Coast Grocery of Tacoma, in which St. Paul & Tacoma had considerable stock. One of Jones's bookkeepers during this period was Peter B. Kyne, who drew on his Grays Harbor experiences in creating the characters and situations in many of his short stories about Cappy Ricks, Matt Peasley, and Tugboat Annie.[3]

During the war, North Western was selected by the Spruce Production Division as one of the mills to cut airplane lumber. When the plant burned on May 22, 1918, the fire was attributed to arson by German agents who were said to have planned the destruction of seven other mills on the Olympic Peninsula. Jones determined to rebuild immediately and went east to order new machinery. While in Milwaukee he suffered a stroke that left him partially paralyzed, but the mill was back in production within six months after the fire.[4]

Though Jones focused most of his attention on the North Western in Hoquiam, he and his wife maintained their residence at 323 North Fifth, only a block away from the Griggs and Hewitt homes. He had an office in the St. Paul & Tacoma headquarters, which he visited regularly when in town.

In 1913, Dr. Edward H. Todd, who had left Tacoma to become vice-president of Willamette University, was offered the presidency of the financially troubled University of Puget Sound. He came back to Tacoma to discuss the future of the school with business leaders. In an unpublished history of the school, Dr. Todd recalled with amusement and satisfaction his visit to the St. Paul & Tacoma offices.

Among those I visited was Charles H. Jones, to whom I said I had been elected to the presidency of the University and if I accepted I would need both financial and sympathetic backing. He said he would give both. I then went into the adjoining office, occupied by his brother-in-law, Henry Hewitt, Jr., and made a similar statement. In his brusque manner he said, 'You ought never to have left here in the first place, for if you had stayed the University would not be in the shape it is in' and he gave a promise of support if I accepted the position.[5]

Later when the school was planning to move from the site now occupied by Jason Lee Junior High to its present campus, Jones pledged $25,000. At the time of his death in 1922 he had redeemed only $5,000 of that amount. Ground-breaking for the new school was set for May 22, 1923. In choosing that date, President Todd was not unmindful of the fact that it was Mrs. Jones's seventy-eighth birthday and the fifty-first anniversary of her marriage.

On the morning of the twenty-second, Dr. Todd called on the widow to extend a personal invitation to be present at the ground-breaking ceremony. He then brought up the school's critical need for money if the building were to be completed.

"How would fifty thousand dollars do?" she asked.

Todd hesitated, then said that he and the trustees were thinking of naming the new building the C. H. Jones Hall. "It would take at least two hundred thousand to erect the kind of building you would wish to see erected in his memory." That afternoon Dr. Todd presented his executive committee with a pledge signed by Franke Tobey Jones for $180,000 in addition to the $20,000 already promised by her husband.[6]

When the formal ground-breaking took place Mrs. Jones stepped between the handles of a plow with a mint-new share. A hundred feet of rope was fastened to the plow and a delegation of students stood by. After the obligatory speeches and prayer, the students grasped the rope and began to pull. Mrs. Jones steered the plow and the ground was broken.[7] When the building was completed in June of 1925, a tablet to the memory of C. H. Jones was mounted in the entrance. For many years it was a tradition to place a basket of flowers beneath the plaque on May 22.

Franke Tobey Jones was a considerable personage. On her husband's death she assumed the presidency of the North Western Lumber Company, holding it until July 16, 1925, when the minority stockholders contracted to purchase her stock and elected Ralph D. Emerson as president.[8]

Franke Tobey Jones, widow of C. H. Jones, at the ground-breaking ceremony for Tacoma's home for the elderly, Marcy 14, 1924 (Tacoma News Tribune)

Having made a major contribution for the young people of Tacoma with her gift to the college, Mrs. Jones determined to do something for those of her own age group. There had been talk for years of the need in the city for a home where the elderly might live in comfort and dignity. A group of Tacoma women, all members of Chapter C of the P.E.O., had incorporated an Old Peoples Home in 1922 and leased a house at 424 D Street, which they furnished with "treasures" from attics and the spare rooms that every home had in those days. Mrs. Jones felt something more was needed. She appeared at a board meeting and said that if it were acceptable she would give the organization five acres of land that she owned adjoining Point Defiance and build a retirement home for sixty-five persons.

Ground was broken on Bristol Street on March 14, 1924. Heath and Gove, who had been architects for Jones Hall, and J. E. Bonnel, the contractor, also built what the trustees chose to call the Franke Tobey Jones Home. On February 4, 1925 the move was made from the old wooden house on D Street. The home was quickly filled. Mrs. Jones had set sixty-five as the limit of residents she felt might live together and keep the atmosphere of a home rather than an institution.[9] It was stipulated that each resident was on probation for a period, as was the home. Once compatability was established, the resident paid a fee of $1,000, which covered all living expenses and care in illness. In addition, the home received half of the new resident's remaining funds, and half of any income such as pensions, royalties, or bequests that might subsequently accrue. This money went to the endowment fund, the interest of which ran the home.

Mrs. Jones never lived there. She died in her own home, April 25, 1931, at the age of eighty-seven.[10] The Joneses were childless but their memory lives on in the Franke Tobey Jones Home and the C. H. Jones Hall.

Jones Hall was not the only contribution from the St. Paul & Tacoma people to the University of Puget Sound. Leonard Howarth left a $100,000 bequest, which provided funds for the construction, alongside Jones Hall, of Howarth Hall.

Hemlock, the Wonderful Weed

WHEN THE TIMBER cruisers brought west by the St. Paul & Tacoma people in 1888 worked their way up the watersheds of the Puyallup, the Mowich, the Carbon, and the Mashell rivers and their tributary creeks, the Voight, Rushing Water, and Busywild, they paid little attention to the hemlock intermixed with the fir, spruce, and cedar. Nobody did, then. To loggers, hemlock was a weed, at best of low commercial value, at worst an awful nuisance.

This disfavor did not result from association with "Socrates drinking the hemlock." Lumbermen literate enough to know the reference also knew that the Greek's suicide draught was brewed from a plant of the genus *Conium,* which was not related to the Amerasian genus *Tsuga.* Western hemlock's bad reputation came partly from the pitchy, unpaintable, sapwood qualities of eastern hemlock, which simply did not make good lumber. *Tsuga heterophylla* had some real problems of its own. It was much heavier than fir, and more expensive to transport. The trees were smaller than fir and produced less clear lumber per log. Hemlock seedlings were shade tolerant; thriving in murk, they tended to crowd out Douglas fir, which needed a bit of sun on its plume.

Early lumbermen could afford to be choosey. Though hemlock was hard to overlook, timber cruisers tended to ignore it, sometimes with the deliberate intent of raising the value of the land on which it grew.

The fairy godfather who liberated this Cinderella tree from the forest scullery was the research chemist. In 1909 the sulphate process for making newsprint out of wood pulp was patented. At first hemlock would not serve as base, but scientists intensified their study of its chemical composition and early in the 1920s found a way to take hemlock apart and reassemble some of it as paper. It was like finding a whole new forest. The discovery was most rewarding to the timber companies that, like St. Paul & Tacoma, had held onto much of their cutover land after the first logging.[1]

When hemlock turned from nuisance to a sought-after raw material, there was a lot of it growing on St. Paul & Tacoma land. Major Griggs in 1927 engaged Norman Porteous and Company, a nationally respected firm of timber cruisers and engineers, to recruise its holdings and give an impartial estimate of what was growing there. At the same time the Western Forestry and Conservation Association was commissioned to make a study of the log production from the original yield and to classify the company holdings of first growth and cutover land for quantity and quality of timber.[2]

While these surveys were in progress—but after it had become apparent that the company owned even more hemlock than had been anticipated —Mayor Mel Tennant brought to Everett Griggs's office a representative of the Union Bag and Paper Corporation of New York. He was seeking a site for a kraft pulp mill. To help Tacoma get a new payroll would St. Paul & Tacoma consider selling the land north of Eleventh Street and west of the Puyallup—the toe of what once had been called "The Boot"?

After protracted negotiations, Leonard Howarth brought the matter to a head. He moved that St. Paul & Tacoma sell the bag company the land for a pulp mill, and that St. Paul & Tacoma then build a new sawmill of its own, one that would specialize in cutting hemlock. Waste from this mill would be sold to Union Bag and Paper as raw material for pulp.[3]

Union Bag agreed to pay $145,220 for 26.41 acres. Three weeks after the sale, the board approved Earl Rogers' design for Mill D, which he estimated would cost $166,000.[4] For Rogers, the decision meant a chance to design a mill from scratch rather than to rebuild an old one as he had done with mills B and C. It was his last such job.

When the mill was complete, a visitor rhapsodized that it was "a child of industry and individual experience, conceded by informed observers to rank in the forefront of modern sawmills." Much of the equipment was original in construction or adaptation, worked out by Rogers and his assistants. Lumbermen were especially impressed by the new log turner, a combination of the usual turner with an electric-driven nigger, the pair controlled by a single lever. Other equipment included a 9-foot Diamond-head mill; an "air-dog, one-man carriage with 14-inch shotgun feed"; a 12-by-72-inch edger; a 7-foot upright Sumner resaw; a 6-foot Allis–Chalmers horizontal resaw; an 8-inch pony edger; two 40-foot and one 30-foot Sumner trimmers; one 40-foot and one 30-foot slasher.

With the addition of Mill D, St. Paul & Tacoma had an eight-hour cutting capacity of 800,000 board feet. Union Bag required 280 cords of wood each day for its pulp operation, of which no more than 35 could be fir. This necessitated putting Mill D on double shift and it raised overall production to more than a million feet a day. The company payroll passed the thousand mark. Hemlock gave Tacoma an expanded payroll at St. Paul & Tacoma, a new payroll at Union Bag, and a new aroma, which reminded some of rotten eggs and others of dollars.

The company bought another 15,000 acres of timberland in the Nooksack Valley. The logs from this purchase were dumped into the bay at Bellingham and rafted south to Tacoma. Closer to home the logging railroad had been ex-

Waste from the St. Paul & Tacoma hemlock mill was hauled to Union Bag (forerunner of St. Regis) to make pulp
(Marvin D. Boland, photographer, 1927; Tacoma Public Library)

233   *Hemlock, the Wonderful Weed*

Medium-sized logs in the St. Paul & Tacoma holding pond (photograph by Marvin D. Boland; courtesy of Wash-ington State Historical Society)

tended to new stands of virgin timber in southeastern Pierce County. Five camps were in operation and each day some 125 carloads of logs rolled down from the mountains to the ponds on the tideflats.

The 200-acre campus was threaded with seven miles of standard-gauge railroad track and forty miles of narrow gauge for the in-plant movement of lumber. Three electric locomotives and a big steam lokey shuffled lumber from green chain to kilns, sheds, planing mills, yards, and docks. There were twenty-four kilns now, all equipped with motor-driven stackers and unstackers, and gadgets that regulated heat, humidity, and air flow. The entire plant was sprinklered and an arsenal of fire-fighting equipment was at hand. There was room to store fifty million board feet of lumber, half of it under roof.

This was the climax of St. Paul & Tacoma's greatest period of production. A reporter visiting Tacoma wrote:

Lumber from the veteran operation shelters the cattleman of South America, the adventurer of the far north, the diamond miner of Africa. Peoples of the Orient eat and drink under its protection. It enshrines family life in Australia. Train passengers in Great Britain and Europe and in remote Siberia ride on rails supported by it. Keels that have carried it across oceans came out of St. Paul trees; as has material for uncounted thousands of freight cars that have transported it over continents. Barges that have landed it at distant wharves and far up strange rivers are built of plank ripped from logs by St. Paul saws. . . .

This native Washington lumber operation has outlived, under one continuous ownership, all of its early contemporaries in the fir regions of the state. When it started, fir and cedar lumber and shingles were not known east of the Rockies. Masts, spars, timbers, large clears and like specialties, so unique and yielded in such abundance from northwest forests, were all that could be sold at first. Much common side lumber, utterly unavailable to any then existing market, was perforce burned up with the refuse of the mill. The products of the operation had to blaze their own trails; and, in doing this have helped to open a sales highway for the whole Pacific northwest lumber industry.[5]

But the year was 1929. The great Wall Street crash had occurred on October 29, the week that this panegyric was published. Years of depression and war lay ahead.

Years of Change

IN 1929, 656 mills in Washington cut 7,302,063,000 board feet of lumber. Came the Great Depression. In 1932, the 333 mills still operating cut 2,260,689,000 board feet. During the same period the average price per thousand board feet fell from $20.59 to $10.73. Roughly half as many mills were cutting a third as much lumber, which sold at half the previous price per unit.

As in the Panic of 1893, St. Paul & Tacoma kept running, but only part time, at reduced pay, and with no profit. The sheds and open storage areas were stacked to overflowing with merchandise that could not be moved. The company joined the Puget Sound Associated Mills, which tried to rationalize shipments to the Atlantic Coast by handling sales without need for middlemen and pooling shipments. Considering the gravity of the situation, this was Band-Aid treatment for a deep wound.

During the depression of the 1890s, Colonel Griggs had been hopeful about the proposed Central Lumber Company, a combination of major producers on the coast which would be able to regulate production and prevent substantial price cutting. The idea was abandoned when some of its proponents tried to create a patently illegal trust. But the concept of a merger of major producers to get the benefits of large operating units would not die.

In 1927, Major Griggs had hosted a meeting in the St. Paul & Tacoma offices at which Alex Polson of Grays Harbor, William Butler of Everett, Mark Reed of Shelton, and other lumbermen, discussed merger, but nothing came of it. Later talks involved possible Weyerhaeuser participation but, as Minot Davis of Weyerhaeuser summed it up, "It is quite evident that a mill merger cannot be formed successfully until such time as mill owners are right up against a necessity of doing something."[1]

Such a time had come by February 1931 when major figures in the industry again gathered in Tacoma to discuss possible amalgamations, which it was claimed "would permit us to improve the product and the service which we can give our customers, and they will afford better protection to employees and stockholders." The mergers under discussion would have resulted in western Washington having only two major companies, Weyerhaeuser and a new giant to be known as the Washington Timber Company, in which St.

Paul & Tacoma would have been a part. Consultations continued into the summer of 1932 but no satisfactory balance of interest could be achieved. Nor was it certain the move could be made without running afoul of the anti-trust laws.[2]

The inauguration of President Roosevelt in 1933 brought massive federal intervention in the economy. One of the first New Deal experiments affecting the lumber industry was the passage of the National Industrial Recovery Act of 1933, which called for industry self-regulation and mandated that codes of fair competition—which would protect consumers, competitors, and employees—were to be drafted for various industries. The NRA was empowered to approve voluntary agreements within an industry to regulate hours of work, rates of pay, and to fix prices. Employees were guaranteed the right to organize and bargain collectively, and could not be compelled, as a condition of employment, to join or refrain from joining a labor organization.

This was a drastic change in the rules. What the government had forbidden in the matter of fixing prices it now approved. The National Lumber Manufacturers Association and subsidiary forest products associations met in Chicago to work out a code, though many of the lumbermen had doubts about giving the government a say in decisions about privately owned forest lands and their harvest.

There was prolonged sectional in-fighting at the code meeting in Chicago in July, especially with regard to wage and price differential between the high-cost western manufacturers and the low-cost mills of the south, but a code maintaining a reasonable balance between the production and consumption of timber products was agreed upon. A twenty-member Lumber Code Authority, drawn from the trade associations, was assigned to calculate every three months the estimated consumption of products for each timber species. The Authority was to supervise wages and labor conditions, log cutting and forestry matters, and handle complaints of violations.

It didn't work. Enforcement was toothless. Lumbermen rebelled against the paperwork and the presence of inspectors. The code did not differentiate between companies like St. Paul & Tacoma, which was set up for continuous operation involving the integrated work of several mills on a single campus producing a variety of types of lumber, and some small outfit which benefited by intermittent cutting when the price was favorable. Inspectors, often brought in from different regions, made decisions that were sometimes incomprehensible and at other times were suspect as putting their home areas in better competitive positions. There was enough chiseling on the codes to make everyone suspicious.[3]

When the West Coast Lumbermen's Association Trustees met in Portland in September 1934 to review the workings of the code, suspicions hardened into hostility. There were demands for an end to the minimum price scale, demands for real punishment of violators, claims that the breakdown in price regulation was "the direct fault of members of the industry," claims that the code was "a canker sore of dishonesty in our industry." Argument became so vehement that J. H. Bloedel, president of Bloedel–Donovan, told his colleagues they sounded like a longshoremen's convention.

The WCLA trustees voted two to one to work for the end of the price protection part of the code, and followed that up with a resolution denouncing the minimum price as "impractical, unworkable and unfair." The WCLA president (Ernest Demerest) and its secretary-manager (W. B. Greeley) submitted letters of resignation in protest of the association's about-face—and the president's resignation was accepted. Legal tests of the validity of code rulings resulted in a bewildering pattern of decisions (price protection was held by district courts to be legal on one side of the Columbia, illegal on the other) until, finally, the air was cleared in 1935 by the unequivocal ruling of the Supreme Court that the National Industrial Recovery Act was unconstitutional.

Tragedy limited the role played by St. Paul & Tacoma in the industry's attempt to make self-regulation work under the NRA.

During the early years of the depression, the Hewitt stockholders had sold most of their shares to obtain capital desperately needed to bolster other family ventures, especially the Hewitt Land Company, which had serious tax problems. The Griggs interest dominated the company, with Everett the central figure, and his brother Herbert serving as vice-president. Their problems were enormous. In December 1931 the National Bank of St. Paul informed them it could not grant the company additional loans. The Bank of California agreed to advance funds to pay 1930 taxes. The following year, St. Paul & Tacoma cut timber at a loss to raise cash for taxes, but still had to borrow more. D. J. Young, the assistant manager of the Bank of California, was put on the St. Paul & Tacoma board, and the bank took over management of the company's indebtedness, which, by July of 1933, rose to $750,000, then began a steady decline as the economy strengthened.

On August 18, 1933, the car Everett Griggs was driving skidded off the road a mile from Everett and struck a tree. Herbert Stanton Griggs died two days later at the age of seventy-two.[4] Everett, who was seventy-one, recovered slowly and never resumed active control of the company. He was succeeded as president by his nephew and namesake, thirty-eight-year-old Everett Gallup Griggs II, the son of C. Milton Griggs, the Colonel's oldest son, who had stayed in St. Paul. This second Everett Griggs was usually referred to as Spike. Another of the Colonel's grandsons, G. Corydon ("Cordy") Wagner, Dr. George Wagner's son, became vice-president. Spike and Cordy worked as a team, with Spike specializing in production, Cordy in sales and industrial relations, but neither had firmly established his authority at the time of the NRA and their immediate concern was the business crisis.

One of the heritages of the NRA was the turmoil in the woods and mills as unions began to organize under the guarantee in Section 7A of the law that employees had "the right to organize and bargain collectively through representatives of their own choosing."

The Loyal Legion of Loggers and Lumbermen was the principal labor organization in the lumber industry in the West when 7A went into effect. It was active at St. Paul & Tacoma and had been looked on with favor by Major Griggs, who had been associated with Colonel Disque at the birth of the organization. But the 4-L, with its strong management representation, was regarded by many workers as a company union, incapable of defending employees' rights. By the time many operators decided the 4-L would be the least of evils, the work force was turning to rival organizations. The American Federation of Labor chartered the Sawmill and Timberworkers Union, while the National Union Unity League, a Communist organization, sponsored the National Lumber Workers Union.[5]

The spring of 1935 brought the great confrontation. In March, the Northwest Council of Sawmill and Timberworkers met in convention at Aberdeen. They accepted the leadership of A. W. Muir, a member of the executive board of the relatively conservative United Brotherhood of Carpenters and Joiners of America, which the AFL had given jurisdiction over the lumber industry. The Council drew up a list of demands covering the industry in the Northwest. They asked to be made the exclusive bargaining agency in each plant or camp. Other demands included: a six-hour day, thirty-hour week; base wage of seventy-five cents an hour, with provisions for overtime and holiday pay; a seniority

George Corydon (Cordy) Wagner, vice-president of St. Paul & Tacoma, in 1955 (Tacoma News Tribune)

system in hiring and layoffs; no strikes or lockouts for the life of the contract until all mediation and conciliation had failed. May 6 was set as a strike deadline.

Negotiations were complicated by disagreements on all sides. The 4–L claimed that 90 percent of the lumber workers opposed a strike. The Communists, taken by surprise by the rapid growth of Timberworkers membership, disbanded their National Lumber Workers Union and set out to infiltrate their AFL rival. Muir's leadership of the council was challenged by local unionists, including the Timberworkers Council vice-president, Norm Lange of Tacoma, on the grounds that he was an outsider who did not know the territory.[6]

The operators couldn't agree whether to try to make the best deal possible with one of the unions or to refuse negotiations while challenging the law in court. The WCLA came out in adamant opposition to a ruling that held that the majority labor group, as determined in an election, should represent all workers in collective bargaining. "At no time and under no circumstance," it said, "will the west coast lumber industry submit to ruling based upon the principle involving denial of representation of any minority groups or employees in collective bargaining." The WCLA subsequently refused to accept conciliation service by the Department of Labor on the ground that "the industry is not organized to deal with its workers as a whole."[7]

Some companies came to terms with locals before the deadline. Others reached agreements with union leaders only to have them rejected by the membership. The McCormick Lumber Company, an important element in the industry, broke ranks and accepted an agreement granting the Timberworkers recognition as the bargaining agent, a modified system of seniority rights, and a forty-hour week with base pay of fifty cents an hour. The strike formally began May 6. Within a week 90 percent of all operations in the Douglas fir region were shut down, St. Paul & Tacoma among them.

A further tangle was added to the snarled situation by the Supreme Court's ruling that the National Industrial Recovery Act was unconstitutional. Though the National Labor Relations Act, which also guaranteed the right to collective bargaining, was on the point of adoption, the court ruling strengthened the operators' inclination to resist, weakened the union position, and led to greater stridency in the speeches of radical leaders as well as a new move by the 4–L to prove the strike unpopular. A postcard ballot that was circulated among employees of Tacoma sawmill, plywood, and door plants drew a 61 percent return. Of those answering, 87 percent said they favored ending the strike on the basis of a 4–L proposal, which called for a modest wage increase. Union officials claimed the results had been rigged, but St. Paul & Tacoma and other Tacoma mills prepared to open behind picket lines.[8]

Spike Griggs led a delegation of employers to conferences with law enforcement officials to ask protection for the plants and workers. Public Safety Commissioner Frank Callender said it would take at least four hundred men to provide security and that he had neither enough policemen available nor funds to hire more. He warned the operators that if the mills resumed operations under armed guard there would probably be a general strike.

Sheriff Jack Bjorklund had been secretary of the Longshoremen's Union before running for office. He told Griggs he would give what protection he could but refused to specify how much that would be. "Start operations and see what happens," he said.

"Will you or will you not take the next step and see the governor?" Spike asked.

"You see him," Bjorklund replied, adding that he thought the operators should agree to collective bargaining and deal with the men. "You're going pretty strong when you ask me to prostitute my principles."[9]

Griggs's delegation found Governor Clarence D. Martin more responsive. He promised that order would be maintained, plants and workers would be protected even if it meant calling out the national guard.[10] The mills re-opened June 21 with state police and deputy sheriffs on guard at the gates. The first day passed quietly, the second brought mass picket lines. Governor Martin called out the National Guard, claiming that Mayor Smitely had requested protection. The Tacoma Labor Council pointed out that there had been no arrests, no disorder, and said that Smitely denied making the request.

The guardsmen and strikers faced each other at the east end of the Eleventh Street bridge on the approach to the St. Paul & Tacoma plant the following morning. Seventy-five soldiers with fixed bayonets moved back a crowd of about 250 pickets who blocked the path of night-shift workers on their way out and day-shift men coming in. There was no violence.[11]

The Central Labor Council called a meeting of all Tacoma unions. A general strike was proposed but not agreed to. The Longshoremen, however, announced that they would not work any ships on Commencement Bay while armed troops were present. Governor Martin reinforced the guard, and their commander issued an order banning mass picketing and prohibiting more than three persons congregating in any one place where troops were present.[12]

Mayor Smitely offered mediation. The union accepted but the operators refused, saying the city had shown that it could not protect the mills and the governor was now in charge. Governor Martin went on the radio to say troops would not be withdrawn until industrial peace was established. "I don't admire or support that employer who would take advantage of the situation to beat down wages," he said, "but I shall continue to demand respect for the right to work."[13]

The federal government, especially concerned that the Longshoremen's tie-up in Tacoma might spread along the coast, intervened. Labor Secretary Frances Perkins, at the request of the United States Conciliation Service, appointed a Portland minister, a Seattle superior court judge, and a Seattle attorney to serve as mediators. The operators claimed "there is nothing to mediate," and protested that the appointment of the board would delay settlement "because naturally the men will hesitate to go back pending the long process of mediation."[14]

On July 11, the city commissioners wired the Federal Mediation Board and the U.S. Department of Conciliation requesting further action. The next day there was real violence. A crowd supporting the workers had gathered outside the post office at Eleventh and A streets. They blocked the approach to the tideflats by parading across the intersection at the sidewalk crossing. The National Guard did not attempt to interfere until 4:30 P.M. let-out time at the mill. Then a line of guardsmen, bayonets at the ready, started marching from the bridge toward A Street, pushing the massed pickets before them. Fights broke out, no shots were fired but one man was clubbed unconscious and tear gas was discharged. Strikers took gas grenades from a National Guard truck and hurled them at the guardsmen, few of whom had masks. The Guard responded by sending a truck, equipped to expel gas from its exhaust, into the area. The truck caught fire and burned. The crowd dispersed.

The arrival in Tacoma of the Federal Mediation Board brought relaxation of tension. A compromise was worked out. In August the employees voted 1,291 to 97 to accept a contract that provided union recognition, an eight-hour day and a forty-hour week, with a base pay of fifty cents an hour.[15]

Labor-management relations at St. Paul & Tacoma, and throughout the lumber industry in the Northwest, had entered a new epoch.

There was bitter irony in another development. The decade that had opened with the government demanding that lumbermen agree to standards and minimum prices ended with the government going to court to charge them with setting prices and using grading standards as part of a conspiracy to limit competition.

From the time it was organized, St. Paul & Tacoma had campaigned for a system of grading lumber to rigid standards. Only so, Colonel Griggs had argued, could Douglas fir break into the midwest market where dealers were accustomed to graded lumber. Lack of standards handicapped producers of quality lumber in competition with ill-equipped mills and damaged the reputation of the industry in the Northwest in their overseas markets. Major Griggs in his role as first president of the National Lumber Manufacturers Association had argued the case for national grades and standards. In 1922, Secretary of Commerce Herbert Hoover spurred the movement with his warning to the lumber associations that if they did not act to establish grading and inspection systems, "we could probably secure the enactment of a 'pure food law' in all building materials."

After prolonged negotiations among the associations, an *American Lumber Standards Handbook* was published, defining the basic grades for softwood lumber and establishing a system of grading and trade marking to show quality.[16] The *Handbook* provoked considerable dissent, especially over the definition that a dressed board of one-inch stock should measure 25/32 of an inch. Many small west coast mills, which lacked dry kilns and shipped their lumber green, could not meet that standard. Secretary Hoover eventually broke the impasse by deciding that 25/32 was standard for dressed one-inch boards, but 26/32 was "extra standard." Later the term became "standard industrial."

With this question of size resolved, the West Coast Lumbermen's Association set up a system of regular grading and inspection. It enforced standards by denying membership to any mill that fell below 95 percent in tests of grading efficiency. This self-policing regulation continued through and after the NRA period.[17]

In 1939, Thurmond Arnold was named to head the antitrust division of the Department of Justice. He announced that one of his first goals would be to "de-NRA-ize" the building industry.[18] The investigation covered a wide range of activities including phony building codes, restrictive agreements between union and contractors, and other practices that could restrain competition. Eventually the FBI accountants began to go through the books of the manufacturers and suppliers of lumber. They found references that they interpreted as indicating collusion among manufacturers to fix prices. The WCLA's inspection system they saw as a possible weapon for punishing those who violated pricing agreements.

In September 1940 a grand jury in Portland returned indictments against five lumbermen's associations, including the WCLA, all of the associations' officers, including President Wagner, sixty-nine corporations, including St. Paul &

Tacoma, and twenty-five individuals. The charges centered on the matter of combining grade marking with trade promotion and shutting out non-members from receiving the benefits of the system.

Nearly all the defendants were indignant at the interpretation put on their actions and inclined to fight the case all the way through the courts, but the outcome was uncertain and the cost in both time and money sure to be high. After months of negotiations it was agreed the defendants would enter a plea of nolo contendere ("I do not wish to contest") and sign consent decrees under which they were under court order to refrain from the alleged monopolistic practices.

A key point in the decree was that the WCLA was not to control the grading and inspection of lumber. The association was ordered to set up a West Coast Bureau of Lumber Grades, which would be managed by an executive committee comprised of both WCLA members and outsiders. The Bureau's inspection services would be available to non-members, who were entitled to stamp their lumber with the approved grade.

The defendants came before the federal court on April 16, 1941, to enter their pleas.[19] Charles Paul, the attorney for the WCLA, told the court: "The defendants are not conscious of any violation of the anti-trust laws. Most business men have been faced with uncertainty about the [Act's] application. . . . Persons named in this indictment, I am sure, have done their best to comply with provisions of the act. We do not admit guilt but in view of the national cri-

Strikers carry flags as they approach national guardsmen blocking their path to the Eleventh Street bridge and the tideflats on July 11, 1935 (Washington State Historical Society)

sis we feel these matters must be settled as they are being settled now."

To the surprise and anguish of the defendants the judge imposed fines of $5,000 each against the associations and smaller fines against officers, trustees, and employees. Fines against the WCLA, its member companies, officers, and employees totalled $64,000. The shareholders voted to pay the fines of its officers, trustees, and employees, and to reimburse member companies on the theory that "it is wholly unjust that these companies alone should pay for the sins of the industry."

Cordy Wagner, speaking for the WCLA, said that the nolo contendere plea was not to be considered an admission of wrong doing. He explained:

It is essentially an agreement to revise business practices in accordance with new rules of the game. The Douglas fir manufacturers have followed the rules of the game as they understood them. The federal courts themselves have been confused as to what the rules are. The government's own economic policy has been uncertain. In the days of the NRA the industry was compelled to do many things, then held to be in the public interest, which are now regarded as grounds for criminal indictment. Industry has had to determine its course in the face of a good deal of uncertainty and confusion. A recent decision of the United

Chief of Police Harold Bird, wearing the white hat, joins in the arrest of a demonstrator who challenged the bayonetted guardsmen (Washington State Historical Society)

States Supreme Court, in the Madison Oil case, seems to extend the Sherman Act to many situations which formerly were not supposed to come under it. The Supreme Court has changed the rules of the game for industry.

We have concluded to play the game in accordance with the new rules. We do this without any consciousness of willful violation of law. The cooperative activities of the West Coast lumber industry, built up around its association, have always been out in the open. Our grade marking, for example, was developed under government encouragement through the Federal Bureau of Standards. Of course the Association had a selfish interest in promoting it: to make West Coast lumber more satisfactory to its users. This was one of the rules of the game, as everyone played it.

Now the Department of Justice regards grade marking as so widespread and important to lumber users that its restriction to members of an association is a monopoly. Like grade marking, our other activities have been carried out in good faith, to meet essential needs of West Coast lumber and without injury to the public. This was true of our industry meetings on trade conditions, of the efforts to keep supply and demand in reasonable balance, of the distribution of differential price lists and recommended shipping weights. We still believe that these services represented business progress; that they benefitted the consumers of lumber as much as the manufacturers.

We shall play the new game in good faith; but we still maintain that nothing was wrong with the old one.[20]

The Renewable Resource

FROM THE FIRST, St. Paul & Tacoma held onto most of its cutover land. Why Chauncey Griggs adopted this practice when it was common practice to let harvested land go on tax-default is not certain. There may be a clue in the fact that much of the white pine area he and Foster logged in Wisconsin became rich farmland. Anyway, taxes were low and there was not much market for stumps. He may simply not have wanted St. Paul's & Tacoma's name on the delinquent tax lists.

Under Everett Griggs's leadership, the company continued to hold onto harvested land. Griggs's emphasis on fire protection gave the land a chance to start natural regeneration. In his presidential address to the 1912 convention of the National Lumber Manufacturers Association, he urged the industry to consider carefully the best uses for logged-off land,[1] but it was not until the 1920s that it became apparent that most of the harvested areas in the Douglas fir region were best suited for growing more conifers.

St. Paul & Tacoma had some 70,000 acres blocked up, guarded against fire, and accessible to roads. Much of it already bore second growth. There was, however, a major problem: taxes.

The State of Washington relied on property taxes for 70 percent of its revenue. Levies increased steadily. They had tripled between 1910 and 1920. Standing timber was assessed higher as it ripened. A section that paid $172 in taxes in 1906 paid $4,100 in 1921. The tax burden encouraged timber owners to cut off trees as quickly as possible, and, with the long growing cycle of Douglas fir, it discouraged systematic reforestation. With taxes, interest, and fire patrol service to be paid each year, costs added to more than the trees were calculated to be worth when ripe. The state encouraged a cut-and-get-out philosophy.

The legislature finally acted to correct the situation. In 1931 it passed the Forest Classification Act, which permitted an owner who registered land as used for timber growing to pay only one dollar an acre while the trees were coming of age but to pay a 15 percent yield tax when they were harvested. With that change the concept of timber as a renewable crop took firm root.

St. Paul & Tacoma had, in 1927, retained Norman Porteous and Company to re-cruise their holdings, and E. T. Allen and Norman Jacobson of the Western Forestry and Conservation Association to make a detailed study of the yield obtained from the original cruise and then to classify the company holdings of timber and cutover land by quantity and quality. Their final written reports, submitted in 1931, served as the basis for a long-range program of reforesting company land, a process St. Paul & Tacoma called tree farming.[2]

The antecedents of the phrase "tree farm" are not certain. George Peavy, dean of the Oregon State College School of Forestry, wrote in a 1928 story in the *4–L News* that "forestry is tree farming." Stewart Holbrook, the logging historian, said in a 1929 editorial in the *Lumber News* that "a logging operator is a tree farmer." Chapin Collins of the *Montesano Vidette* is credited with recommending the use of the term "tree farm" in reference to reforestation projects "because it creates a visual image—everybody knows what a farm is."[3] During the 1930s an American Tree Farm League was formed, and a program established under which any area dedicated by its owner "to the continuous production of forest crops" and certified by an inspecting forester can be designated a tree farm. The name has been given to everything from a one-acre stand to a forest empire such as St. Paul & Tacoma's.

The reforestation program is nourished by non-profit tree nurseries supported by the forest products industry. The first of these nurseries was established in 1941 on the Nisqually flats not far from the mouth of Sequalitchew Creek from which the Hudson's Bay Company shipped the first Puget Sound lumber to market a century earlier. Cordy

Wagner, as president of the West Coast Lumbermen's Association, took the lead in establishing the program and making certain that the Forest Industries Tree Nursery (now called the Col. W. B. Greeley Forest Nursery) was located within sight of the most traveled highway in the state, so the public could see that lumbermen sowed as well as reaped.

Indeed, the nursery was a delight to the eye and a pledge of allegiance to the concept of sustained yield. Driving past one saw acre after acre of seedlings—Douglas fir, Port Orford cedar, Western hemlock, Sitka spruce—cupped between the protecting hills of the Nisqually estuary. The tree farm was originally designed to produce more than six million young trees a year (it has since been greatly expanded) for transplanting on the logged and burned-off lands of the participating companies.

Seeds were gathered from the cones of genetically superior trees. They were planted in broad beds of rich soil, and the seedlings were toughened by controlled periods of drought and wetting. Harvesting came in December and January. The soil around the seedlings was loosened mechanically but each tiny tree, about five inches tall but with a root system half again as long, was lifted by hand and placed on an endless belt in a nearby shed. Nursery workers discarded any seedlings with signs of weakness, wrapped each sturdy one in a layette of sphagnum moss and damp-resistant paper, and packed them six thousand to a crate for shipment to the hills to be reforested. Teams of young foresters carefully tucked them in, eight feet apart, five hundred to the acre, and watched with pride the growth of organisms whose life span exceeded their own.

"Tree farm" as a phrase is public relations. As a fact it has proved good business, and as good business it promises to endure. Clear cutting still leaves the land looking as if mechanized armies had clashed upon it, but one has only to look to the surrounding hills to see the light greens of the next generation rising from the stock started in the seed beds on the Nisqually and its sister nurseries.

Some of the seedlings have gone to Tomalla, a private tree farm established by Spike and Mary Lea Griggs on a township (160-acre) tract just outside Orting. Tomalla is a showcase tree farm, which serves as outdoor classroom for thousands of local school children each year. The Industrial Forestry Association uses a portion of the farm as a seed nursery, where cones are collected for study.

More intensive research is conducted by the Puget Sound branch of the Pacific Northwest Forest and Range Experiment at the Voight Creek Experimental Forest. This 230-acre area, which lies on a slope above the Carbon River, was logged by St. Paul & Tacoma in 1900. It reseeded itself. In 1947 the company dedicated the Voight Creek forest to research on second growth. Foresters have divided the terrain into plots that are thinned at varying rates to determine which produces the maximum growth. Companion studies measure seed production under partially controlled conditions.

Research has become the growing tip of the industry.

Only Change Goes On Forever . . .

THE SECOND World War brought less change to St. Paul & Tacoma operations than had the First. Shipbuilding resumed on the tideflats, but aircraft carriers could not be built of fir. Aluminum had replaced spruce as the staple of the aircraft industry. But barracks were still constructed of boards, ordnance was crated for shipment in wooden boxes, and pallets were essential for loading and unloading. To meet the need for such material the company's treasure of virgin timber was drawn down faster than had been planned—a sacrifice to the war effort, and the mill complex on the tideflats ran double shift from Pearl Harbor to V-J Day.

With millions under arms, manpower was in shorter supply than timber. Officers at the Pierce County military establishments cooperated in a program under which off-duty soldiers, sailors, airmen, and coastguardsmen could volunteer for periodic shifts on the green chain. Production goals set by the government were met. The army–navy E for excellence fluttered below the stars and stripes on the spruce pole at the mill gate.

The end of the war marked the end of St. Paul & Tacoma's long and sometimes profitable association with the Wilkeson Coal and Coke Company. Colonel Griggs and Henry Hewitt were no strangers to coal when they came to Tacoma, and were not unmindful of the happy coincidence that underneath part of their Cascade timberlands lay beds of bituminous coal, a heritage from eons past when heavy plant growth laid down beds of vegetable matter that turned to carbon under pressure.

Coal had been discovered in the Cascades east of Puget Sound as soon as Hudson's Bay Company traders began ascending the rivers. The deposits in the canyon of the Carbon River were first reported in the 1860s but not until the Flett brothers of Orting took a location claim on Flett (now Gale) Creek in 1873 was there mining in the drainage system emptying on Commencement Bay. A wagon road was run to the Flett diggings and some coal carted to Tacoma in the same month that the Northern Pacific's spur line from Kalama on the Columbia River reached the town.[1]

The Northern Pacific commissioned an engineer, Benjamin Fallows of Pittsburgh, to investigate coal prospects. When he reported favorably they ran the route for their proposed Cascade Division through the coal area. The Northern Pacific was in financial straits during the period but its president, Charles Wright, put forward personal funds to enable the company to lay track for thirty miles into the mountains east of Tacoma. The town at the railhead was named Wilkeson for Sam Wilkeson, longtime secretary of the NP board of directors and a leading publicist of Puget Sound resources. The original Wilkeson mine, worked by Tacoma Coal and Coke, was abandoned and a second started on the western dip in 1879.[2]

A new company, Wilkeson Coal and Coke, began operations on the Smith vein in the eastern dip around 1880. It was one of the first companies in Washington Territory to manufacture industrial coke. At first they distilled the bituminous coal by burning it in rude pits, but soon they built coking ovens that resembled huge beehives.

In 1890 St. Paul & Tacoma bought a controlling interest in Wilkeson Coal and Coke. The company owned 50 percent of the stock and individual officers held most of the remaining shares.[3] Wilkeson built 40 more beehives (eventually there were 160) and signed a long term contract to supply high-grade metallurgical coke to the Tacoma smelter. Other major customers for coke were the Tacoma Gas Company and the Seattle Gas Company. Hundreds of thousands of tons of what was considered the best bituminous coal in the Cascades was delivered to the gravity bunkers that the Northern Pacific built on the Tacoma waterfront below Cliff Avenue (Stadium Way), making Tacoma the best place on the coast for steamers to bunker.

Immigrants with mining experience, most of them from Wales, Italy, or the Balkans, found their way to Wilkeson where they cut coal in the shafts or quarried sandstone that was used for many of the public buildings in Washington, including the capitol. Wilkeson was a typical miners' town of the period, radical in politics and rough in entertainment. There were thirteen saloons and only one bank. License fees on the drinking spots paid the entire cost of maintaining the streets until prohibition bankrupted the town's treasury, moonshiners not paying taxes.

That was the way things went in Wilkeson, bust following boom, strike following speed-up. Production peaked during the First World War, then fell off rapidly as diesel replaced steam in the merchant marine and navy, and electricity and oil supplanted coke in the smelting process. The mines were closed during a national coal strike in the thirties and never came back to full production. Mining stopped altogether in 1936; most of the colliers moved away; a 1941 guidebook described Wilkeson as "a ghost community . . . the present population for the most part dependant on relief." After Pearl Harbor, with the nation at war and needing coal, the town came back to life.

St. Paul & Tacoma joined with the Northern Pacific in establishing the Wilkeson Products Company; its mission was to reopen the mines and build a coke plant in Tacoma.

The War Production Board committed $700,000 of the estimated $900,000 capital requirement. Cordy Wagner was president of Wilkeson Products. Chauncey Leavenworth Griggs, son of Herbert Griggs & grandson of Colonel Griggs, served as secretary and as project expeditor, coordinating activities of the railroad and the lumber company with those of government agencies.[4]

Chauncey L. Griggs, after graduating from Yale with a degree in French literature, entered the family lumber business. While working as a tally man in the sawmill, he took courses in cost accounting at Tacoma's Knapp Business College night school. These studies left him with a life-long enthusiasm for the study of balance sheets and the drafting of memoranda interpreting their import.

Appointed assistant secretary of St. Paul & Tacoma in 1932, Chauncey Griggs was elected to the board on his father's death, and served as secretary. He also became director and vice-president of the Alaska Transportation Company, a corporation established by S. A. Perkins and Norton Clapp, in which St. Paul & Tacoma held a small interest. When the Coastal Steamship Company was formed by St. Paul & Tacoma to operate a former Great Lakes freighter, the *Lake Frances,* between Tacoma and California, he was appointed president and manager. In a private venture, he was founder and president of Ski Lifts, Inc., which built and operated the first ski tows west of New England.[5]

The Wilkeson resurrection did not give the coal and coke company eternal life. There were wrangles on the board over the best methods of working the seams, which were broken by severe faulting. Griggs resigned from the company in 1943 and joined the army. The coke production enterprise was liquidated in 1944[6] and Wilkeson went back to sleep, perhaps to dream of a new awakening in a time of high demand and improved technology.[7]

St. Paul & Tacoma's involvement in the operation of the *Lake Frances* ended more happily. When the Coastal Steamship Company was incorporated in 1936, $30,000 of its $50,000 capitalization represented the book value of the ship. The *Lake Frances* carried lumber to San Pedro for the Tacoma Lumber Sales Company, which had been created by a consortium of Tacoma mills to serve the Los Angeles area. The ship returned with general cargo for Puget Sound ports.

When lumber sales slowed in 1938, the Coastal directors instructed Griggs to try to find a buyer for the ship. L. C. Hammond of San Francisco, chairman of Hammond Lumber, offered $40,000. Griggs and Dickman recommended against acceptance, arguing that if Hitler plunged Europe into war, any ship that would float would increase in value. Coastal turned down Hammond's offer and chartered out the vessel for runs between California and the Caribbean. After the war started rates went as high as $400 a day, until the *Lake Frances* was requisitioned by the government to carry supplies to Alaska, the Aleutians, and the South Pacific.

When the war ended, the *Lake Frances* was sent to Oakland to be reconditioned for return to her owners. Chauncey Griggs, who happened to be stationed in San Francisco, was able to monitor the work done on her and exultantly assured the firm that the government was turning back the *Lake Frances* in better shape than when she was drafted. Coastal put her up for sale—she is thought to have been the first ship sold after the war—and the Lee Chee Company of China bought her for $250,000 cash, more than six times the offer made by Hammond before the war.[8]

Virgin stands of western hemlock and Douglas fir had become rare in the 1950s (from West Coast Lumberman)

The Final Chapter

PHILOSOPHERS OF BUSINESS theorize that a family-owned company must suffer disagreements between members engaged in active management, who want to plow profits back into expanded facilities, and the more removed shareholding relatives whose concern is the size of the dividends. Integrated companies, which produce the raw materials they later process, face differences between the divisions supplying basic ingredients and those responsible for manufacturing goods and getting them to market. St. Paul & Tacoma, a family-owned, tree-to-warehouse operation, in the last years of its corporate existence was spared neither type of conflict.

The reorganization of the board that brought Spike Griggs and Cordy Wagner to control in 1933 came at the bottom of the great depression. The question for St. Paul & Tacoma at the time was not the distribution of profits; the question was survival. Hard times made dissent a luxury few permitted themselves. Leaders set direction; everyone pulled on his oar. Recovery in the late thirties was followed almost immediately by the crisis of World War II. The problems of wartime production were those of shortages,

shipping, regulations, government specifications, and labor; of demands imposed by an expanded bureaucracy, which, though necessary and inevitable, was often inexperienced and exasperating. Some disputes arose within management about interpretations of directives and regulations but they did not lead to any overall challenge of the leadership. Spike and Cordy shared responsibility, though the latter played the more active role in formulating policy.[1]

Post-war conditions, however, revealed and exacerbated differences in interest and philosophy among the board members. The conflict grew out of the question of whether St. Paul & Tacoma should manufacture lumber or concentrate on managing its forest heritage. This was a re-run of the old Hewitt–Griggs dispute, but with a changed line-up.

In meeting the challenges the depression created for their other interests, the Hewitts had been forced to sell much of their St. Paul & Tacoma stock. The Hewitt family, which now held only about 8 percent of the shares, was represented on the board by Henry Hewitt IV. He personally held few shares but was active in management, having taken over the distribution yard, long a losing proposition, and turned it into one of the company's most profitable operations.

The Howarth interest in the company had risen to nearly 33 percent. These shares were held by the three daughters of William Howarth; they had inherited not only their father's holdings but those of their bachelor uncle Leonard. William Pilz and Dr. A. H. Meadowcroft, both married to Howarth daughters, were on the board. The Griggs interest, constituting 53 percent, a clear majority, was represented by Spike, Cordy, and Chauncey. The remaining 6 percent of the shares was scattered.

Ben Grosscup, the company's attorney, was the only non-family member on the board, but he was allied with the Griggses, which left control of the company and its policy with Colonel Griggs's descendants, as long as there was no split in family ranks.

In March of 1950, Secretary Chauncey Griggs submitted a memorandum to Spike, Cordy, and "our associates on the board," in which he challenged the basic assumptions under which the company (and most other large timberholding companies) was operating.[2] Citing figures from the annual reports, he argued that had it not been for the rise in value of its old-growth timber, St. Paul & Tacoma between 1920 and 1939 would have suffered losses greater than all its earnings plus the original $1,500,000 investment. "Our company," said the secretary, "has survived in the past by reason of unearned appreciation in the value of timber."

The memo went on to quote figures from industry-wide reports on 1948 operations which, according to Griggs, indicated St. Paul & Tacoma's logging costs were 48 percent higher than the average among competitors with similar rail-and-cargo operations; the manufacturing costs 73 percent higher. He concluded that the manufacturing end of the operations was being subsidized by setting a less-than-market price on logs delivered to the company mills from St. Paul & Tacoma timberlands. Estimating the loss at "between $1,200,000 and $1,800,000 a year, depending on whether the true market value is placed at three or four times book value," he called for "an impartial, thorough, over-all survey of our operations by a first class firm of management engineers."

Cordy responded crisply with an inter-office note telling Chauncey "You have another think coming, if you expect me to analyze and check the chart which you state has taken a great deal of your time and so far has been criticized by us with silence. . . . I do not expect to attempt to seriously digest it until many other more pressing matters are out of the way. We are spending our time for such studies in preparing forecasts of the year ahead, and changes in operation which will correct many things which have given concern this last year."[3]

This drew another memo from Griggs: "The point is that we are kidding ourselves if we think it is only costing us $3.92 more than the average combination mill to take logs at the dump at 'market prices,' make them into lumber and sell it." He marshalled figures supporting his contention that St. Paul & Tacoma's cost of manufacturing was higher by $11.87 per thousand board feet than for similar combination mills reporting to the West Coast Lumbermen's Association.

Wagner fired off a letter to the entire board which said that while Griggs's figures might have "some theoretical interest" they were, "for any practical purpose, valueless."[4] He pointed out that St. Paul & Tacoma had been erroneously listed among the combination mills in the WCLA report for 1948. The association considered combination mills to be those that shipped from 35 to 65 percent of their lumber by water, but St. Paul & Tacoma that year had put less than 10 percent of its cut on ships.

Furthermore, the Tacoma mill had been working mainly in hemlock, white fir, and low-grade Douglas fir. The company's high-grade fir was used for plywood. To compare a mill cutting hemlock with one cutting prime fir was to compare apples and oranges. He deplored the whole discussion, which "has diverted attention from particular operations where we can and must make some savings in payroll and/or expense, to some quite theoretical computations. The main thing is to lower man-hour costs per thousand feet, both in the woods and at the mill, and also to raise the profit realization from our products wherever possible."

At the next annual meeting, a motion by H. Stanton Griggs, Chauncey's brother, for a management survey was defeated by the stockholders.[5] Cordy went ahead with his plans to diversify the product and lower man-hour costs through product expansion and plant modernization.

Immediately after the war, St. Paul & Tacoma had installed new planers in the main planing mill, increased shed storage, and added dry kilns with automatic stackers. More significantly, they had purchased the Olympia Veneer Company plant on Budd Inlet. Olympia Veneer was the pioneer plywood co-operative, founded in 1921 by 125 investors who managed to raise $1,000 apiece, then work in the mill for half a year without pay while perfecting their process of peeling huge logs toward their core and glueing the sheets together, crosswise, to form a material stronger than natural wood. The co-op was so successful that several of its owner-workers financed plants of their own.[6]

When the use of plywood in the paraphernalia of World War II—PT boats, assault craft, floating dry docks, Quonset huts—proved to a doubting public that plywood sheets no longer curled back like dandelion stems when wet, demand increased. The problem was that plywood required old-growth timber, of which there was only so much to be purchased. Olympia Veneer owned no stumpage of its own. Deciding to move south to Oregon, where peeler logs cost less on the open market, they sold the mill on Budd Inlet to St. Paul & Tacoma, which had old fir to peel. Acquisition of the plywood plant led to the diversion of much of St. Paul & Tacoma's old-growth fir to Olympia, and to the Tacoma mill's concentration on hemlock and white fir.[7]

The pulp industry, too, offered opportunity for diversifi-

A peeler block being unwound for plywood in Tacoma about 1940 (Mary Randlett copied this photo by an unknown photographer from a print in the Tacoma Public Library)

cation in the use of the company's timber holdings. In 1950 Cordy Wagner won board approval for installation of a barker-chipper at Mill C. On being pulled from the storage ponds, each log was directed to the barking chamber. Steel clamps gripped each end and rotated the brown carcass. An operator screened by shatterproof glass directed against its turning side a jet of water that struck with a force of 1,400 pounds per square inch. There was a scream of liquid against solid, a fog of flying fiber, a pause as debris settled. Ninety seconds after entering the chamber the log, skinned to its yellow flesh, moved on toward the headrig while the peeled particles of its side flowed to a conveyor and were taken to a grinder, which reduced them to hogged fuel and dressing for planter beds.

As the naked log was carried against the band saws, its curved outer slabs dropped onto endless belts that moved them to chipping machines. They were chopped into bark-free flakes small enough for digestion in the maw of a pulp mill. St. Paul & Tacoma sold its chips to the St. Regis Kraft Company, a wholly-owned subsidiary of the St. Regis Paper Company, one of the giants of the pulp industry.

The St. Regis pulp mill was the one Union Bag had built in 1928 at the toe of the boot, on land purchased from St. Paul & Tacoma and the foresightful Leonard Howarth. Union Bag went pop during the depression. St. Regis bought the mill, left it idle between 1932 and 1936, then reopened it with new machinery, a bleaching plant, and a furnace that reduced the kraft odor of rotting eggs by a degree perhaps perceptible to sophisticated instrumentation.[8]

Diversification of product and the increased use of each log meant greater need for logs. It was the quest for additional stumpage that brought St. Paul & Tacoma into association with Hilding Lindberg, a self-made timber baronet.

Hilding was the son of Gustaf Lindberg, a Scandinavian immigrant who had built up a wholesale grocery business that specialized in serving logging camps. Gustaf branched into various ventures which climaxed with his founding the ill-fated Scandinavian–American bank. It went bust with an embarrassing bang. The skeleton of the Washington Building, of which Scandinavian–American was to have been the prime tenant, cast a ghastly shadow over downtown Tacoma for several years.

As a young man Hilding had sold groceries to the camps in the tall timber. Left on his own after his father's failure, he organized a company to operate cook houses and mess halls for logging companies cutting in the distant foothills. The time was right. The Wagner Act of 1935 spurred union organization; better living conditions in the camps were high on the list of worker demands. Through volume buying and standardized operations, Lindberg not only met the demands of loggers' palates and stomachs, he cut cook-house costs.

His trips to outback camps gave Lindberg a fresh perspective. He noted that easily harvestable timber was running out, and he foresaw an era when larger and more powerful trucks, and improved dozers, would make it economical to run logging roads up slopes too steep for locomotives. Aided by loans from Philip Weyerhaeuser and other lumbermen he had met while peddling provender, he formed a timber company and began to pick up high-altitude stumpage, timber just beyond the fringe of technology and profit in the thirties.

After the war, well after the start of the era of truck logging he had foreseen, Lindberg faced what he feared was a terminal illness. He decided to dispose of his timber holdings. While negotiating to sell the bulk of his timber to Georgia Pacific, he disposed of a tract adjoining St. Paul & Tacoma holdings in Lewis County to the Tacoma firm for $100,000. This sparked the interest of Cordy and Spike.[9]

Chauncey Griggs recalled receiving a telephone call from Spike saying a meeting of the board was to be held in Cyrus Happy's law office in the Rust Building and asking him to come immediately. He met Spike in the hall outside the office. "Spike asked if I would be in favor of spending something like $4,500,000 to pick up a block of old growth timber in the company's operating area at $18 per thousand. That was indeed a bargain price. I was operating tie mills at the time and we were paying almost that much for second growth in patches of not over half a million feet. It didn't take me more than twenty seconds to agree. I asked Spike who in the world would offer such a big block at that bargain price. 'Hilding Lindberg,' he said. 'I think he's going to buy the Pilz stock and we'll probably put him on the board.' "

The William J. Pilz stock, roughly one-third of the total Howarth holding, had been on the market for about a year with no takers. Mr. Pilz was dying of cancer and was anxious to get his affairs in order. Chauncey Griggs reasoned that Cordy Wagner, who wanted the shares to end up in friendly hands, saw Hilding as an ally. Hilding was feeling better, doubted that he would die soon, and would prefer selling to a company that would put him on the board instead of selling to one where he would receive nothing more than money out of the deal. Chauncey still felt he had been correct in his 1950 memos. Knowing Hilding by reputation as "a genuine 100 percent money-maker," he assumed that after looking at the operating figures Hilding's conclusions would match his own.[10]

The deal went through. Hilding sold his timber holdings to St. Paul & Tacoma for $5,000,000. Henry Hewitt, Jr., the company treasurer, wrote out the check. Lindberg attended the meeting in logging clothes and a secretary present mused afterwards, "I just saw an old bum walk out of the office with a check for five million bucks in the pocket of his wool shirt."[11]

Lindberg used the money to buy the Pilz stock. He was elected to the board, the first true outsider to sit on it, and was permitted, even encouraged, to buy other odds and ends of stock that came up for sale. A forceful man with ideas and ambitions of his own, Lindberg soon came to have considerable influence with Spike, a situation that alarmed some other board members who came to look on him as conniving and ruthless.

Henry Hewitt, Jr., recalled being warned by a Tacoma banker that "Hilding is playing too close to the railroad tracks," a reference to Lindberg's supposed ambition to become a director of the Northern Pacific. According to Chauncey Griggs, "No businessman in his right mind accepts such a position in a vast, cumbersome organization which is actually operated by various division managers miles and miles below the Board, whose members are in essence figureheads. Hilding kept his eye on the ball: make money for Hilding. Let the boys who hanker for such empty positions occupy them. What Hilding was after in his flirtations with Northern Pacific was a hold on their remaining vast reserves of old-growth timber available to St. Paul sawmill operations. If he had pulled this off, it might have totally changed the final outcome. All I know for sure is that almost as soon as he became a St. Paul & Tacoma board member he was talking about the possibility of tying into this NP timber for the company. A seat on the NP board would have been a nuisance, not a help, in this effort."[12]

Whatever Lindberg's private thoughts, he won Spike Griggs's acceptance of the idea of replacing Attorney Ben Grosscup on the board with Anson Moody. There were good arguments for the move. Moody was the husband of the third Howarth daughter; his election would return the Howarth interest on the board to what it had been before Lindberg acquired the Pilz stock. Further, Moody was the long-time manager of Everett Pulp & Paper, and St. Paul was considering building a pulp mill.

Cordy Wagner in his search for diversification and greater utilization of the company's timber resources, had decided there would be more profit in manufacturing pulp than in selling chips to St. Regis. Pulp mills were extremely expensive, and the trend in the industry was toward ever larger plants. Western Kraft in Portland, however, had recently designed and constructed a 100-ton plant that was making money. After consultations with Western Kraft's executive officer and chief engineer, Wagner proposed that a similar plant on the Tacoma tideflats would provide a relatively inexpensive way for St. Paul & Tacoma to enter the pulp industry.[13]

Lindberg and Moody both opposed the idea and joined Chauncey Griggs in contending that the company's forest base was not adequate to support a pulping operation. Lindberg favored expanding timber holdings through purchase or by entering an alliance with one of the major timber owners, perhaps Northern Pacific. Moody objected that a 100-ton plant would be too small even if it could be built at the estimated cost.[14] Wagner argued that the plant would be ideal for St. Paul & Tacoma's available timber, but faced with resistance pursued alternative possibilities for diversified operations. One involved Lindberg's idea of association with the Northern Pacific in the creation of a jointly owned forest corridor feeding logs to the Tacoma mill. Another, quickly dropped, was a sustained-yield agreement with the government, similar to the Simpson hundred-year-cycle on the Olympic Peninsula. The most promising was an arrangement under which St. Regis would spin off its Tacoma pulp mill into a separate organization, which would then merge with St. Paul & Tacoma.

While seeking some arrangement with other operators, Wagner found himself in a surprise battle for control of St. Paul & Tacoma. The Griggs's interest still represented 53 percent of the total shares; the Howarths had 26 percent, Lindberg, 10 percent, the Hewitts, 8 percent, with the remaining 3 percent scattered. The Griggs votes would have been enough to keep Cordy in firm control had the Griggs's block been solid. It proved not to be.

Lindberg was pursuing an independent course. Chauncey Griggs believed that the newcomer, after studying the books, had indeed arrived at the same conclusions that Chauncey had presented in his 1950 memo. Lindberg later told him of confronting Spike and Cordy at a private meeting and arguing unsuccessfully about operating costs. He worried that most of his wealth was locked into a company he now thought was dissipating its resources at a dangerous rate.[15]

Henry Hewitt, Jr., who supported Wagner, remembered a meeting with Lindberg at which he rejected an offer of the presidency of the company if he would join the opposition.

Hewitt turned it down and then had to fight off attempts to force him to resign from the board.[16]

One day in August, 1955, Chauncey received a phone call from Lindberg asking for a meeting at the New Yorker Restaurant on Sixth Avenue. "It was the first direct communication I ever had from him." As Chauncey sat down in a booth in the coffee shop, Lindberg asked, "How are you going to get me out of this fix?"[17]

The key to the power struggle lay in the C. W. Griggs Investment Company, which had been formed not long before Colonel Griggs's death to serve as "an instrument for holding, perpetuating and expanding the family interest." When the Colonel died in 1910, the shares were divided equally among his six children. The Investment Company's original holding of St. Paul & Tacoma stock amounted to 4,249 shares. These were augmented in 1914 by the purchase of the 1,252 Foster shares, and subsequent additions had raised the Investment Company's total to 7,154 of the 15,000 shares issued. It was Investment Company policy to allow Cordy Wagner to vote its shares as a block.

The family holdings were no longer equal. Two of the six families—the Anna Griggs Tilton family and the C. Milton Griggs family—had each disposed of approximately 30 percent of their shares to the other four families. As a result, the shares held by the Herbert S. Griggs family and the Theodore W. Griggs family, both of which supported Chauncey's position, totaled 827.25. The C. Milton Griggs shares, which supported Spike Griggs's position, and the

Anna Griggs Tilton shares, which supported Cordy Wagner, totaled 990.

The control of the Investment Company rested with Major Griggs's widow, Grace Wallace Griggs, who held 280.25 shares, and her position was enigmatic. Reviewing the situation with Lindberg, Chauncey pointed out that his family's stock, combined with that of Mrs. Theodore Griggs, and any one of the other four blocks would be sufficient to control the Investment Company. He suggested that Lindberg, Anson Moody, and Dr. Meadowcroft, husband of the third Howarth daughter, should be the ones to talk to Grace Griggs.[18]

The next day Chauncey flew to New York. On the plane he drafted a proposal for a new pooling agreement, under which those who signed would agree to put their shares in the Investment Company into the hands of three trustees for a seven-year period, the shares to be voted as a block in accordance with the decisions of the trustee majority. Chauncey proposed that the three trustees be Spike Griggs, Anson Moody, and Hilding Lindberg.

Mrs. Theodore Griggs was the first to sign the pooling agreement. Next Chauncey, his sister, and his brother signed in Tacoma.[19] The decision rested with Grace Wallace Griggs. Moody, Lindberg, and Dr. Meadowcroft each visited her in San Francisco. She finally agreed, signing at a conference attended by her attorney, Owen Hughes, and Moody and Lindberg.[20] Her decision may have been influenced by her nephew, Craig Wallace, whom Mrs. Griggs had often appointed to represent her at conferences where differences of opinion surfaced. (It was later learned that she had already written a will leaving her entire estate to her brother's children, the Wallaces.)

Whatever her reasons for signing, her shares gave the insurgents 1,110 votes; the Cordy and Spike management team had only 990. The coup was a *fait accompli*. At this point Spike agreed to serve as a member of the trustee committee, bringing with him the C. M. Griggs family shares, which were helpful but not necessary for voting control of the Investment Company stock.[21]

In January, 1956, the insurgents presented Cordy with a letter which said, "People generally hesitate to become involved in situations where a minority interest can exert voting strength greatly exceeding its actual holdings." They asked that the Investment Company distribute its St. Paul & Tacoma shares to the stockholders.[22] With the trustees, Spike Griggs, Lindberg, and Moody, in position to vote a majority of the Investment Company shares, there was no argument. The Investment Company was liquidated.

Cordy Wagner made one last counter-attack. He moved that the Lumber Company board be enlarged to twelve members, but the proposal lost, 5 to 2, with Henry Hewitt, Jr., casting the only vote supporting Cordy. The insurgents were in firm control. No changes were made in officers of the company, but power had shifted.

Negotiations continued with St. Regis. As discussions went on the objective shifted from that of securing for St.

City Waterway and the south shore of what once was "The Boot" when the St. Paul and Tacoma Lumber Company was absorbed by St. Regis (Harry Boersig, photographer; author's collection)

Paul & Tacoma a proprietary interest as principal or partner in a pulp operation to that of arranging a merger in which St. Paul & Tacoma would be absorbed by the larger company.

After several near breakdowns in the talks, St. Regis, early in 1957, offered an exchange of stock under which St. Paul & Tacoma shareholders would receive 56⅔ shares of St. Regis for each share of St. Paul. The transfer was to be made across a three-year period, with one-third of the stock transferred each first of August in 1959, 1960, and 1961.

Cordy Wagner, believing that the company was just beginning to gain the full benefit of some $20,000,000 invested in capital improvement and timber acquisition in the decade following the war, demurred. Henry Hewitt, Jr., proposed further talks with the Mead Corporation, the nation's fifth largest paper manufacturer, and the exploration of any promising alternatives. The majority voted for acceptance. They cited capital gains benefits, a more favorable estate tax situation for elderly shareholders, insulation from the risks involved with the approaching exhaustion of the company's old-growth timber, and the probable beneficial effects on St. Regis stock when the pulp company acquired 133,700 acres of first- and second-growth Douglas fir and hemlock within forty miles of its Tacoma plant. The stockholders voted to accept the St. Regis tender.[23]

There was sadness in Tacoma at the disappearance of the name St. Paul & Tacoma. The creation of the company in 1888 had spurred the greatest boom in the city's history. The steadiness of its management during the Panic of 1893 when fourteen banks, the Northern Pacific and the Tacoma Land Company all went into bankruptcy, held the framework of community together, though people left town by the thousands. In the following years, St. Paul & Tacoma supplied not only a payroll but often political leadership. To many in the city, St. Paul & Tacoma seemed as much a part of their lives as the mountain, the bay, and the tideflats.

The mill remains. St. Regis, through the merger, was able to expand operations in the forest. Henry Hewitt, Jr., provided some continuity, staying with St. Regis until 1979 when his retirement ended ninety-one years of continuous executive connection with the operation by members of the founding families.

The vapor plumes of the pulp mill rise high over the tideflats and the former St. Paul & Tacoma mill still stands on "The Boot," which the company transformed into one of the major manufacturing areas of the Pacific Coast.

Notes

June 4, 1888

1. Material in this chapter is drawn largely from the minutes of June 4, 1888, St. Paul & Tacoma Lumber Co. Records, University of Washington Libraries; see also *Tacoma Ledger,* June 5, 1888, and *Tacoma News,* June 5, 1888.

Founding Fathers

1. The sketch of Chauncey Wright Griggs is drawn from C. A. Snowden, *History of Washington: The Rise and Progress of an American State* (New York: Century History Co., 6 vols., 1909), 5:58–63; and from the following subscription histories: Julian Hawthorne, *History of Washington, the Evergreen State* (New York: American Historical Publishing Co., 2 vols., 1893), 1:408–12; H. K. Hines, *An Illustrated History of the State of Washington* (Chicago: Lewis Publishing Co., 1893), pp. 247–50; Herbert Hunt and Floyd Kaylor, *Washington, West of the Cascades* (Chicago: S. J. Clarke Publishing Co., 1917), pp. 128–30; Herbert Hunt, *Tacoma, Its History and Its Builders* (Chicago: S. J. Clarke Publishing Co., 3 vols., 1916), 3:10–15; and William F. Prosser, *A History of the Puget Sound Country* (New York: Lewis Publishing Co., 2 vols., 1903), 1:310–13. Also consulted were the Meany Pioneer File and the Dubar scrapbooks in the Northwest Collection, University of Washington Libraries.

2. Harriet Bishop, *Floral Home; or, First Years in Minnesota* (New York, 1857), p. 125 (cited in Albro Martin, *James J. Hill and the Opening of the Northwest* [New York: Oxford University Press, 1976], p. 29).

3. "The Narrative of the Third Regiment," by General C. C. Andrews, is from *Minnesota in the Civil and Indian Wars, 1861–65,* published under the supervision of the Board of Commissioners appointed by the Act of the Legislature of Minnesota, April 16, 1889 (St. Paul, Minn., 1890), pp. 145–66.

4. The best account of the Hill–Griggs partnership is in Martin, *James J. Hill and the Opening of the Northwest.*

5. Articles of co-partnership between Hill, Griggs, and William B. Newcomb, August 20, 1869, James J. Hill Papers, General Correspondence, Minnesota Historical Society. Attached to the agreements of January 1, 1872, formalizing the Red River enterprise, are agreements of August 2, 1870, and February 20, 1871 (cited in Martin, *James J. Hill and the Opening of the Northwest*).

6. Diary, Hill Papers, May 5, 1873 (cited in Martin, *James J. Hill and the Opening of the Northwest,* p. 98).

7. C. M. Underhill to Hill, September 20, 1875, Hill Papers, General Correspondence, Minnesota Historical Society.

8. Hill to Griggs, May 30, 1876; Hill to Underhill, November 19, 1875; Underhill to Hill, December 2, 1975 (cited in

Martin, *James J. Hill and the Opening of the Northwest,* pp. 106–7).

9. The sketch of Addison Foster is drawn from *Legislative Manual of Washington State, 1899; Washington Standard* (Olympia), February 3, 1899; *Biographic Dictionary of the American Congress* (Washington, D.C., 1971), p. 962; *National Cyclopaedia of American Biography* (New York, 1904), 12:390; "Personal Recollections of Everett Worthington Foster," manuscript in Minnesota Historical Society archives; autobiographical letter, Everett Foster to Minnesota Historical Society, May 28, 1928.

10. *St. Paul Daily Globe,* December 3, 1878.

11. Hines, *Illustrated History of the State of Washington,* pp. 248–49.

12. Ralph W. Hidy, Frank E. Hill, and Allan Nevins, *Timber and Men: The Weyerhaeuser Story* (New York: Macmillan Co., 1963), pp. 111–12.

13. Griggs to Henry Dimmock, October 12, 1898. Chauncey Griggs Letterpress copy books, St. Paul & Tacoma Lumber Co. Records, University of Washington Libraries.

14. A. G. Harvey, *Douglas of the Fir* (Cambridge: Harvard University Press, 1947), pp. 57–58 (cites David Douglas' "Journal," p. 339).

15. Sketch of Henry Hewitt, Jr., is drawn from William Whitfield, *History of Snohomish County, Washington* (Chicago: Pioneer Historical Co., 1926), pp. 309–51; Snowden, *History of Washington,* 5:166–71; Hawthorne, *History of Washington, the Evergreen State,* pp. 526–28; Hines, *Illustrated History of the State of Washington,* pp. 526–29; Hunt and Kaylor, *Washington, West of the Cascades,* pp. 391–93; Hunt, *Tacoma, Its History and Its Builders,* 3:30–36; *Tacoma Sunday Ledger,* October 3, 1909; obituary, *Seattle Post-Intelligencer,* May 2, 1918.

Biographies of Henry Hewitt, Sr., are in Publius V. Lawson, *History of Winnebago County, Wisconsin* (Chicago, 1908), 2:781–

84; and *United States Biographical Dictionary: Portrait Gallery of Eminent and Self-Made Men* (Chicago, 1877), pp. 458–59.

16. *Oshkosh Times,* May 8, 1888 (reprinted in *Tacoma Ledger,* May 19, 1888).

17. John Bell Sanborn, "The Story of the Fox–Wisconsin Rivers Improvement," in *Proceedings of the State Historical Society of Wisconsin,* 47th annual meeting, Green Bay, Wisconsin, September 5–7, 1899.

18. Lawrence University Records, 1853–54; Marguerite Ellen Schumann, *Creation of a Campus* (Appleton, Wis., 1957), pp. 6–9; *Lawrence Yearbook,* 1853.

19. Morgan L. Martin Papers, University of Wisconsin Library.

20. Information on the business activities of both father and son is contained in the Mercantile Agency reports for Neenah–Menasha, 1848–84, in the Dun and Bradstreet Collection, Wisconsin State Historical Society.

21. Lawson, *History of Winnebago County,* p. 56.

22. John Strange, "Autobiography of John Strange," Day 22, 1922, manuscript, John Strange Papers, Wisconsin State Historical Society.

23. *Appleton (Wis.) Crescent,* June 21, 1856; *Green Bay Advocate,* June 22, 1856.

24. Sanborn, "The Story of the Fox-Wisconsin Rivers Improvement," pp. 193–94; "Law and the Promotion of Enterprise," pp. 598–600, Samuel Mermin Papers, Wisconsin State Historical Society.

25. *Tacoma Sunday Ledger,* October 3, 1909.

26. "Bank of Menasha's Colorful History," *Twin City News Record,* August 14, 1963; *U.S. Biographical Dictionary,* p. 458; Lawson, *History of Winnebago County,* pp. 781–82.

27. Strange, "Autobiography," pp. 10, 12a.

28. "Sketch of the History of First National Bank, Appleton, Wisconsin," privately printed pamphlet, courtesy of the research department, First National Bank, Appleton; "Menasha Wooden Ware Company History," anonymous typescript, Local History File, Menasha Public Library.

29. Lawson, *History of Winnebago County*, 2:1011.

30. *Neenah Daily Times*, October 10, 12, 1883.

31. Sketch of Charles Hebard Jones is drawn from Snowden, *History of Washington*, 5:171–75; W. P. Bonney, *History of Pierce County, Washington* (Chicago: Pioneer Historical Publishing Co., 3 vols., 1927), 3:34–36; Hawthorne, *History of Washington, the Evergreen State*, 2:526–29; Hunt and Kaylor, *Washington, West of the Cascades*, 3:391–94; Hunt, *Tacoma, Its History and Its Builders*, 3:30–34; Prosser, *History of the Puget Sound Country*, 1:298–301; *Encyclopedia of American Biography*, 12:346; "Menominee, the Gem of the Upper Peninsula: The Historical and Industrial Record of the Great Lumber Center," Meany Pioneer File, University of Washington Libraries, December 1899, p. 52.

32. "Catalogue of the Officers and Students of Ripon College for the Collegiate Year 1865–66" (Charles H. Jones, Menasha, is included with the "Gentlemen Students Preparing for College").

33. "The Great Peshtigo Fire," pamphlet, Peshtigo Historical Society.

34. "Menominee, the Gem of the Upper Peninsula," p. 52.

35. Ethel Schuyler, ed., *Menominee County Book for Schools* (Menominee, 1941), p. 158.

36. Alvah L. Sawyer, *A History of the Northern Peninsula of Michigan and Its People* (Chicago, 1911), 2:575.

37. "Menominee, the Gem of the Upper Peninsula," p. 52.

38. The Nogales Chamber of Commerce reports that a small smelter operated briefly in the city in the 1880s. They have no specific information about it.

The Big Deal

1. Background for the discussion of freight rates is to be found in Thomas R. Cox, *Mills and Markets: A History of the Pacific Coast Lumber Industry to 1900* (Seattle: University of Washington Press, 1974).

2. Oakes's proposal is based on a story in the *Tacoma Ledger*, May 3, 1888.

3. Herbert Hunt, *Tacoma, Its History and Its Builders* (Chicago: S. J. Clarke Publishing Co., 3 vols., 1916), 3:32.

4. Norman J. Johnston, "Frederick Law Olmsted and the Plan for Tacoma," *Pacific Northwest Quarterly*, 66 (July 1975):97–104.

5. Abstract of contract, St. Paul & Tacoma Lumber Co. Records, University of Washington Libraries. Northern Pacific board minutes, March 20, 1888, pp. 181–82; September 19, 1888, pp. 240–41, in File 3E10 7B, Northern Pacific Papers, Minnesota Historical Society.

6. *Oshkosh Times*, May 8, 1888 (reprinted in *Tacoma Ledger*, May 19, 1888; *Pioneer Press* (St. Paul, Minn.), June 1, 1888 (reprinted in *Northwest Magazine*, July 1888).

A Mill Is Born

1. The sketch of George Brown is draw from C. A. Snowden, *History of Washington: The Rise and Progress of an American State* (New York: Century History Co., 6 vols., 1909), 5:164–66; William Prosser, *A History of the Puget Sound Country* (New York: Lewis Publishing Co., 2 vols., 1903, 1:496–98; *Sketches of Washingtonians* (Seattle: W. C. Wolfe and Co., 1906), pp. 122–23.

2. Randolph Foster Radebaugh, "Memoirs," typescript, nos. 31, 32, 33, Tacoma Public Library.

3. Coast and Geodetic Survey charts 1463 and 5690, Commencement Bay.

4. *Tacoma News,* June 9, 1888, p. 8.

5. *Tacoma News,* June 16, 1888, p. 4; June 18, 1888, p. 4.

6. *Tacoma News,* June 9, 1888, p. 4.

7. *Tacoma Ledger,* April 23, 1889, p. 1.

8. Radebaugh, "Memoirs," no. 40, p. 5.

9. Minutes (microfilm), St. Paul & Tacoma Lumber Co. Records, University of Washington Libraries.

10. Obituary of P. D. Norton, St. Paul & Tacoma Lumber Co. Records, University of Washington Libraries.

11. *Tacoma Ledger,* June 10, 1888, p. 4.

12. *Tacoma Globe,* April 14, 1889, p. 3.

13. *Tacoma News,* April 6, 1889, p. 1.

14. *Tacoma Ledger,* April 16, 1889, p. 5; April 23, 1889, p. 1.

"Now You're Logging . . ."

1. *Tacoma News,* April 6, 1889, p. 1.

2. Thomas Cox, *Mills and Markets* (Seattle: University of Washington Press, 1974), chap. 9.

3. *West Shore,* 1882, p. 619.

4. Cox, *Mills and Markets,* chap. 9.

5. Bruce B. Cheever, "The Development of Railroads in the State of Washington, 1860 to 1948," Master's thesis, Western Washington College of Education, 1949, pp. 100–101.

6. *Puget Sound Lumberman,* January 1893, pp. 39–40.

7. *In Business to Stay,* pamphlet published by the St. Paul & Tacoma Lumber Company, 1947, p. 6, St. Paul & Tacoma Lumber Co. Records, University of Washington Libraries.

8. *Puget Sound Lumberman,* January 1893, p. 29.

9. Brandon Satterlee, *The Dub of South Burlap* (New York: Exposition Press, 1952), pp. 119–20.

10. *Tacoma News,* April 6, 1889.

To Market We Must Go

1. Griggs to Dimmock, October 22, 1898. Chauncey Griggs Letterpress copy book, St. Paul & Tacoma Lumber Co. Records, University of Washington Libraries.

2. *Tacoma Call,* October 29, 1891.

3. *Tacoma Globe,* November 18, 1891, p. 1.

4. *Puget Sound Lumberman,* January 1893, pp. 39–40.

5. Ibid.

6. Vernon H. Jensen, *Lumber and Labor* (New York: Farrar & Rinehart, 1945), p. 115.

7. *Tacoma Globe,* November 18, 1891, p. 1.

8. *Puget Sound Lumberman,* January 1891; February 1892.

9. *Tacoma Globe,* December 16, 1891, p. 5.

10. Minutes, January 4, 1892, St. Paul & Tacoma Lumber Co. Records, University of Washington Libraries.

11. Minutes, January 14, 1892, St. Paul & Tacoma Lumber Co. Records.

Booming the Boom Town

1. Rudyard Kipling, *From Sea to Sea* (London: Doubleday & McClure, c. 1899).

2. *Tacoma News,* June 15, 1888.

3. Russell Earley, "The St. Paul & Tacoma Lumber Company," unnumbered typescript, St. Paul & Tacoma Lumber Co. Records, University of Washington Libraries.

4. Herbert Hunt, *Tacoma, Its History and Its Builders* (Chicago: S. J. Clarke Publishing Co., 3 vols., 1916), 3:20–25.

5. Randolph Radebaugh, "Memoirs," typescript no. 42, Tacoma Public Library.

6. Ibid., no. 43.

7. Ibid., no. 49.

8. Bruce B. Cheever, "The Development of Railroads in the State of Washington, 1860 to 1948," Master's thesis, Western Washington College of Education, 1949, pp. 68, 100–101.

9. *Tacoma News,* April 19, 1889.

10. Hunt, *Tacoma, Its History and Its Builders,* 3:478–79.

11. Minutes, July 8, 1889, St. Paul & Tacoma Lumber Co. Records, University of Washington Libraries; see also, Thomas Ripley, *Green Timber* (Palo Alto, Calif.: American West Publishing Co., 1968).

12. *Tacoma News,* April 5, 1889.

13. Minutes, May 31 and June 9, 1890, St. Paul & Tacoma Lumber Co. Records.

14. Griggs to Hewitt, April 18, 1898, Chauncey Griggs Letterpress copy books, St. Paul & Tacoma Lumber Co. Records, University of Washington Libraries.

15. Minutes, September 8, 1890, St. Paul & Tacoma Lumber Co. Records.

16. *Tacoma News,* April 6, 1889, p. 1.

17. O. S. Van Olinda, *History of Vashon–Maury Islands* (Vashon, Wash.: Vashon Island News–Record, 1935).

18. *Tacoma Globe,* August 24, 1890, p. 5.

19. Radebaugh, "Memoirs," no. 53.

Putting the Puyallup in Its Place

1. U.S. Congress, House, "Survey of the Mouth of the Puyallup River, Washington," 55th Cong., 2d sess.

2. Washington State Highway Department, "Reconnaissance Geological Report, Tacoma Spur," project 31144-AO, pp. 1–2.

3. Coast and Geodetic Survey, Commencement Bay, chart T-1749, 1886.

4. *Tacoma Globe,* August 7, 1890.

5. Hans Bergman, ed. and pub., *History of Scandinavians in Tacoma and Pierce County* (Tacoma, Wa., 1926), pp. 9–10.

6. *Tacoma Ledger,* May 7, 1889, p. 5; May 20, 1889, p. 4.

7. Material on the flood is drawn from the *Tacoma Globe, Tacoma News,* and *Tacoma Ledger,* November 10–29, 1891.

8. "Survey of the Mouth of the Puyallup River," pp. 5–6; also, minutes, December 10, 1892, St. Paul & Tacoma Lumber Co. Records, University of Washington Libraries.

9. Coast and Geodetic Survey, Commencement Bay, 1892.

10. "Survey of the Mouth of the Puyallup River."

11. The map is with the Washington State Historical Society.

"We Live Here, Too"

1. *Barton's Legislative Handbook, 1889* (Olympia, Washington), p. 374.

2. *Tacoma Globe,* August 7, 1890, p. 3.

3. Snowden, *History of Washington: The Rise and Progress of an American State* (New York: Century History Co., 6 vols, 1909), 5:165.

4. *Everett News,* February 12, 1892, p. 2.

5. *Tacoma Globe,* September 15, 1890, p. 5.

6. *Tacoma Ledger,* April 21, 1889, p. 2.

7. *Tacoma News,* August 13, 1892, p. 7; August 10 and August 16, 1892.

8. *Tacoma News,* June 29, 1892, p. 3.

9. *Barton's Legislative Handbook, 1889,* pp. 376–78.

10. Griggs to James Hamilton Lewis, June 23, 1896. Chauncey Griggs Letterpress copy books, St. Paul & Tacoma Lumber Co. Records, University of Washington Libraries.

11. Herbert Hunt, *Tacoma, Its History and Its Builders* (Chicago: S. J. Clarke Publishing Co., 3 vols., 1916), 2:130–39.

12. Griggs to A. Trowbridge, January 16, 1895, Griggs Letterpress copy books.

13. Griggs to Hewitt, March 29, 1895; Griggs to Isaac W. Anderson, April 8, 1895, Griggs Letterpress copy books.

14. *Puget Sound Lumberman,* September 1892.

15. Hunt, *Tacoma, Its History and Its Builders,* 2:579–80.

16. *Tacoma News,* November 12, 1892, p. 2.

The Everett Connection

1. W. D. C. Spike, *Spike's Illustrated Description of the City of Tacoma* (Tacoma, Wash.: Spike & Co., 1891), unnumbered.

2. "Henry Hewitt Tells Inside Story of Everett's Founding," *Everett Herald,* March 13, 1913, p. 1.

3. Allan Nevins, *Study in Power: John D. Rockefeller, Industrialist and Philanthropist* (New York: Scribner, 2 vols., 1953), 2:203–5; Norman H. Clark, *Mill Town* (Seattle: University of Washington Press, 1970), pp. 21–33.

4. E. W. Wright, ed., *Lewis and Dryden's Marine History of the Pacific Northwest* (Portland: Lewis and Dryden, 1895), p. 386.

5. *Everett Herald,* March 13, 1913, p. 1.

6. Ibid.

7. Clark, *Mill Town,* p. 22.

8. William Whitfield, *History of Snohomish County, Washington* (Chicago: Pioneer Historical Co., 1926), pp. 310, 314–16.

9. Ibid., p. 314.

10. "Francis Brownell Relates Some Little Known Facts about Early Everett, Monte Cristo Mines," *Everett Herald,* Oct. 12, 19, 1974. (Historical file, U.S. Forest Service, Monte Cristo Division.)

11. Griggs to Hewitt, April 16, 1898. Chauncey Griggs Letterpress copy books, St. Paul & Tacoma Lumber Co. Records, University of Washington Libraries.

12. *Spike's Illustrated Description of the City of Tacoma.*

13. Whitfield, *History of Snohomish County,* p. 317.

14. W. P. A., *The New Washington,* 1941, pp. 186–200; Whitfield, *History of Snohomish County,* p. 331.

15. Norman Clark, "The Men Who Staked Millions on 'Pittsburgh of Puget Sound,'" *Seattle Post-Intelligencer,* December 13, 1970.

16. *Everett Herald,* March 13, 1913, p. 1.

17. Quoted in Whitfield, *History of Snohomish County,* pp. 227–28.

18. *Everett Herald,* March 13, 1913, p. 1.

19. Whitfield, *History of Snohomish County,* pp. 339–40.

20. Wright, ed., *Lewis and Dryden's Marine History,* p. 386; Whitfield, *History of Snohomish County,* pp. 328, 339; *Everett News,* October 21, November 27, December 22, 1891 (all p. 1); December 25, 1891, p. 8; January 29, 1892, p. 1.

21. *Everett News,* February 19, 1892; Clark, *Mill Town,* pp. 10–17; *Everett Times,* February 25, 1892.

22. *Everett News,* November 29, 1891, p. 4.

23. *Monte Cristo Mountaineer,* July 1, 1964; Rosemary Wilkie, *A Broad Bold Ledge of Gold: Historical Facts of Monte Cristo, Washington* (Seattle: Seattle Printing and Publishing, 1958).

24. Marshall Bond, Jr., *Gold Hunter: The Adventures of Marshall Bond* (Albuquerque: University of New Mexico Press, 1969), pp. 7–11.

25. *Seattle Post-Intelligencer,* June 12, 1890.

26. Thomas Burke to Mary M. Miller, September 25, 1891 (quoted in Robert Nesbit, *"He Built Seattle"* [Seattle: University of Washington Press, 1961], p. 270).

27. "Francis Brownell Relates Some Little Known Facts About Early Everett," *Everett Herald* (undated clipping).

28. *Everett News,* March 12, 1892.

29. Whitfield, *History of Snohomish County,* pp. 341–42.

30. Clark, *Mill Town,* pp. 20–27.

31. Ibid., p. 28.

32. Albro Martin, *James J. Hill and the Opening of the Northwest* (New York: Oxford University Press, 1976), p. 396.

33. Clark, *Mill Town,* p. 29.

34. *Everett Herald,* March 13, 1913, p. 1.

35. Francis Brownell, "Why the Monte Cristo Mines Were a Mistake," *Everett Herald,* undated clipping.

36. Ralph W. Hidy, Frank E. Hill, and Allan Nevins, *Timber and Men: The Weyerhaeuser Story* (New York: Macmillan Co., 1963), pp. 209–10.

37. *Tacoma Ledger,* October 3, 1909.

38. A letter from Martha Griggs to Morgan, October 5, 1979, states that "an interview with Mr. Rockefeller is not specifically mentioned in the records and a summary of any agreement reached is not extant." Rockefeller Archive Center.

39. Gates to Crocker, February 2, 1895, Letterbook 332, p. 265; Gates to W. W. Huntington, August 14, 1895, Letterbook 333, p. 235, Rockefeller Archive Center.

40. Whitfield, *History of Snohomish County,* pp. 350–51.

41. Griggs to E. H. Bailey, February 11, 1895, Griggs Letterpress copy books.

Crash and Crisis

1. Minutes, January 14, 1892, St. Paul & Tacoma Lumber Co. Records, University of Washington Libraries.

2. *Puget Sound Lumberman,* April 1892; March 1892.

3. Ibid., May 1892.

4. *Tacoma Globe,* May 18, 1892.

5. *Puget Sound Lumberman,* June 1892.

6. Ibid., August 1892; Griggs to Hewitt, July 18, 1892, Chauncey Griggs Letterpress copy books, St. Paul & Tacoma Lumber Co. Records, University of Washington Libraries. (Subsequent correspondence in this chapter is also found in the Griggs Letterpress copy books.)

7. *Puget Sound Lumberman,* December 1892.

8. *West Coast Lumberman,* February 1893.

9. Ibid., March 1893; July 1893.

10. Ibid., August 1893.

11. A. D. Noyes, *Thirty Years of American Finance* (New York and London: G. P. Putnam's Sons, 1902), pp. 190–94; Matthew Josephson, *The Politicos, 1865–1896* (New York: Harcourt, Brace and Co., 1938), pp. 526–31. Minutes, June 12, 1893, St. Paul & Tacoma Lumber Co. Records; Northern Pacific board minutes, June 21, 1893.

12. *West Coast Lumberman,* May 1894.

13. Ibid., April 1894.

14. *Tacoma Evening Call,* October 16, 1891.

15. Herbert Hunt, *Tacoma, Its History and Its Builders* (Chicago: S. J. Clarke Publishing Co., 3 vols., 1916), 2:110–15; *Tacoma Ledger* and *Tacoma News,* July 18–23, 1893.

16. Griggs to George Browne, January 12, 1894.

17. *Tacoma Ledger,* January 25, 1894; *Tacoma News,* January 24, 25, 1894.

18. *Tacoma Ledger,* May 20, 1894.

19. Griggs to A. Trowbridge, October 18, 1894; January 16, 1895.

20. Griggs to Trowbridge, October 18, 1894; Griggs to C. M. Griggs, October 30, 1894.

21. Griggs to Theodore Griggs, March 6, 1894; April 11, 1894; November 11, 1894; October 4, 1895; March 4, 1896.

22. Griggs to Hiram F. Garretson, July 25, 1895.

23. Griggs to Leroy Palmer, October 23, 1894.

24. Griggs to J. W. Cooper, July 20, 1896; Griggs to James J. Hill, July 17, 1897.

25. Griggs to Ambrose Tighe, January 5, 1895; Griggs to E. H. Bailey, January 5, 1895; Griggs to Jes C. Eckels, January 7, 1895.

26. Griggs to Isaac Anderson, January 25, 1895.

27. Griggs to Henry Hewitt, March 29, 1895; Griggs to Addison Foster, August 27 and December 3, 1895; Griggs to M. M. Ladd, June 28, 1898; Griggs to A. Trowbridge, June 13, 1898.

28. E. W. Wright, ed., *Lewis and Dryden's Marine History of the Pacific Northwest* (Portland, Ore.: Lewis and Dryden, 1895), p. 403. Griggs to W. B. Blackwell, April 11, 1894; Griggs to H. W. Lamberton, April 16, 1894; Griggs to T. H. Titus, April 16, 1894; Griggs to A. Trowbridge, August 16, 1894; Griggs to C. H. Jones, February 19, 1895; Griggs to Capt. Charles Nelson, February 19, 1895; Griggs to Henry A. Strong, May 7 and October 8, 1898.

29. Griggs to A. Trowbridge, April 30, 1895.

30. Griggs to Addison Foster, August 26, October 1, and November 2, 1895; Griggs to Henry Hewitt, September 23, 1895; Griggs to C. M. Griggs, February 10, 1896; Griggs to Foster, May 4, 1896.

31. Griggs to C. M. Griggs, July 6 and 20, 1896.

32. Griggs to Senator John L. Wilson, November 18, 1896.

33. Griggs to Mark Hanna, November 16, 1896.

34. Griggs to A. Trowbridge, January 27, 1897.

35. Griggs to Henry Dimmock, October 22, 1898.

New Markets

1. *West Coast Lumberman,* October 1896.

2. Griggs to Addison Foster, September 30, 1896. Chauncey Griggs Letterpress copy books, St. Paul & Tacoma Lumber Co. Records, University of Washington Libraries. (Subsequent correspondence in this chapter is also found in the Griggs Letterpress copy books.)

3. *West Coast Lumberman,* January 1897.

4. Griggs to S. L. Levy, August 3, 1897.

5. James N. Tattersall, "The Economic Development of the Pacific Northwest to 1920," Ph.D. dissertation, University of Washington, 1960, pp. 137–40; Tattersall, "Exports and Economic Growth: The Pacific Northwest, 1880 to 1960," *Papers and Proceedings of the Regional Science Association* (1962), pp. 215–34.

6. Thomas Cox, *Mills and Markets* (Seattle: University of Washington Press, 1974), pp. 200, 205, 287–88; *Tacoma Ledger,* August 11, 1898.

7. Capt. Harold D. Huycke, "E. R. Sterling," *Oceans* 7 (September 1974):38; "Mighty Ship Was the *Everett G. Griggs,*" *Tacoma Ledger,* December 13, 1964.

8. Griggs to Hattie M. Lockwood, March 4, 1898; Griggs to Roxana Wilson, March 14, 1898; Griggs to Anna B. Griggs, April 12, 1898; Griggs to Henry Hewitt, Jr., April 18, 1898; Griggs to Capt. Charles Nelson, June 25, 1898; interview with Corydon B. Wagner, April 7, 1979.

9. Griggs to Hewitt, April 18, 1898.

10. Griggs to Anna B. Griggs, June 20, 1898.

11. Griggs to Theodore W. Griggs, April 22 and June 1, 1898.

12. Griggs to Anna Griggs, July 11, 1898; Griggs to C. M. Griggs, July 19, 1898.

13. *West Coast Lumberman,* June 1900.

A Senator from St. Paul and Tacoma

1. The political background of this period in Washington is well summarized in Robert Nesbit, *"He Built Seattle": A Biography of Judge Thomas Burke* (Seattle: University of Washington Press, 1961).

2. Extracts from Addison Foster's diary from 1884 to 1899 are in a manuscript entitled "A Saga of 70 Years," prepared by Russell Earley for Corydon Wagner, Jr. (Hereafter referred to as Foster diary; the original is the property of Charles A. Foster, Jr.)

3. *Tacoma Ledger,* August 11, 1898.

4. *Tacoma News,* August 11, 1898.

5. *Mississippi Valley Lumberman,* 30 (5).

6. Interview with Mrs. William Pettit (Foster's granddaughter), of Lexington, Kentucky.

7. Foster diary, October 6, 1898.

8. *Tacoma Ledger,* October 7, 1898; *Tacoma News,* October 7, 1898.

9. *Puget Sound Lumberman,* January 1899.

10. *Tacoma Ledger,* January 10, 1899.

11. *Seattle Times, January 12, 1899.*

12. *Spokane Outburst,* quoted in *Tacoma Herald,* July 28, 1894; see also *Tacoma Herald,* July 7, 1894.

13. *Whatcom Blade,* quoted in *Tacoma Herald,* December 1, 1894.

14. Albert Johnson "Life in Tacoma in 1898," published in *Tacoma News-Tribune,* July 4, 1934. The most detailed account of the voting in the caucus and the legislature is found in the *Morning Olympian,* January 25–February 3, 1899.

15. Foster diary, January 30, 1899.

16. Ibid., February 1, 1899.

17. *Washington Standard,* February 3, 1899.

18. *Mississippi Valley Lumberman,* February 3, 1899.

19. Foster diary. Foster's record in the Senate was one of automatic support of the national party position. He paid scant attention to the party in the state and was not a serious contender for renomination after his freshman term.

"The Most Perfect Sawmill"

1. Griggs to Henry Hewitt, Jr., April 18, 1898. Chauncey Griggs Letterpress copy books, St. Paul & Tacoma Lumber Co. Records, University of Washington Libraries.

2. Ibid.

3. *Mississippi Valley Lumberman,* March 1, 1901, pp. 26–28.

4. Ibid.

5. Ibid.

6. *West Coast Lumberman,* December 1893.

7. *Mississippi Valley Lumberman,* March 1, 1901, p. 26.

Organization Man

1. Stewart H. Holbrook, *Burning an Empire: The Story of American Forest Fires* (New York: Macmillan Co., 1943), pp. 108–20; Ellis Lucia, *Head Rig: The Story of the West Coast Lumber Industry* (Portland, Ore.: Overland West Press, 1965), pp. 49–51; William B. Greeley, *Forests and Men* (New York: Doubleday, 1951), pp. 19–20.

2. Herbert Hunt, *Tacoma, Its History and Its Builders* (Chicago: S. J. Clarke Publishing Co., 3 vols., 1916), 3:15–17.

3. Griggs to Foster, November 9, 1895. Chauncey Griggs Letterpress copy books, St. Paul & Tacoma Lumber Co. Records, University of Washington Libraries.

4. *Tacoma Globe,* November 18, 1891; John H. Cox, "Trade Associations in the Lumber Industry of the Pacific Northwest, 1899–1914," *Pacific Northwest Quarterly* 41 (October 1950):285–311; Vernon H. Jensen, *Lumber and Labor* (New York: Farrar & Rinehart, 1945), p. 115; *West Coast Lumberman* October 1891, p. 3; December 1891, p. 1; February 1892, p. 9.

5. Griggs to James Hamilton Lewis, March 4, 1897. Griggs Letterpress copy books.

6. Thomas Cox, *Mills and Markets* (Seattle: University of Washington Press, 1974), pp. 289–90.

7. Lucia, *Head Rig,* pp. 50–51; Cox, *Mills and Markets,* pp. 309–10.

8. Lucia, *Head Rig,* p. 51; Ralph Hidy, Frank Hill, and Allan Nevins, *Timber and Men: The Weyerhaeuser Story* (New York: Macmillan Co., 1963), pp. 229–31, 375–76, 381, 385.

9. Greeley, *Forests and Men,* p. 20.

10. Hidy et al., *Timber and Men,* p. 242.

11. Lucia, *Head Rig,* p. 52.

12. Hidy et al., *Timber and Men,* p. 243.

13. Ibid., p. 242; Cox, "Trade Associations in the Lumber Industry," p. 310.

14. Greeley, *Forests and Men,* p. 20.

15. Lucia, *Head Rig,* p. 52.

16. Ibid., p. 51.

17. Cox, "Trade Associations in the Lumber Industry," pp. 307–8.

18. Ibid., p. 311.

19. *Seattle Star,* June 29, 1907 (Joe Smith).

20. Ibid., June 30, 1907.

21. Ibid.

22. Cox, "Trade Associations in the Lumber Industry," p. 305.

23. Ibid., p. 306.

24. Lucia, *Head Rig,* p. 58; Hidy et al., *Timber and Men,* pp. 234–35; E. T. Coman, *Time, Tide and Timber: A Century of Pope & Talbot* (Stanford: Stanford University Press, 1949), p. 223.

25. Hidy et al., *Timber and Men,* p. 235.

26. Lucia, *Head Rig,* pp. 61–62; Cox "Trade Associations in the Lumber Industry," p. 302–5.

27. Cox, "Trade Associations in the Lumber Industry," p. 305.

28. Ibid., pp. 291–92; Lucia, *Head Rig,* pp. 62–63; *American Lumberman,* December 11, 1903, p. 32; *Pacific Lumber Trade Journal,* April 11, 1911.

29. Department of Commerce and Labor, Bureau of Corporations, *The Lumber Industry* (Washington, D.C.: U.S. Government Printing Office, 1913), pp. 12–13.

30. Ibid., part 4, p. 18.

31. *Pacific Lumber Trade Journal,* April 11, 1911, p. 29.

The Changing of the Guard

1. Minutes, February 12, 1908, St. Paul & Tacoma Lumber Co. Records, University of Washington Libraries.

2. Interview with Corydon Wagner.

3. *Tacoma Ledger,* October 3, 1909, p. 26.

4. Herbert Hunt, *Tacoma, Its History and Its Builders* (Chicago: S. J. Clarke Publishing Co., 3 vols., 1916), 3:30–33.

5. W. C. Pettit, "Northwest Lumber Co.," in *Washington*

State Industries and Resources: Lumber Mill Histories, typescript, Hoquiam Public Library.

6. A Glimpse into the Lives of Chauncey Wright Griggs and His Wife Martha Ann Gallup Griggs. Privately printed, February 1978. Golden Jubilee, 1859–1909, Martha Ann Gallup, Chauncey W. Griggs. Privately printed, 1909.

7. Tacoma Ledger, October 29, 1910; Tacoma News, October 30, 1910. St. Paul Dispatch, October 29, 1910; St. Paul Pioneer Press, October 30, 1910.

8. C. M. Griggs to family, April 16, 1913. Chauncey Griggs Letterpress copy books, St. Paul & Tacoma Lumber Co. Records, University of Washington Libraries.

Disaster and Discord

1. Tacoma News, June 8, 9, 10, 1912; Tacoma Ledger, June 8, 9, 10, 1912; Records, Tacoma Fire Department, June 9, 10, 1912. Minutes, July 8, 1912, St. Paul & Tacoma Lumber Co. Records, University of Washington Libraries.

2. "A Story of the Development of One of America's Greatest Lumber Manufacturing Institutions," American Lumberman (paid advertising supplement), May 21, 1921, pp. 98–100.

3. Minutes, February 10, 1913, St. Paul & Tacoma Lumber Co. Records. The Potlatch mill designed by Wilkinson and Son is described in Ralph Hidy, Frank Hill, and Allan Nevins, Timber and Men: The Weyerhaeuser Story (New York: Macmillan Co., 1963), p. 256.

4. Hidy et al., Timber and Men, p. 274; Norman Clark, Mill Town (Seattle: University of Washington Press, 1970), pp. 128–29.

5. Minutes, February 10, 1913, St. Paul & Tacoma Lumber Co. Records. Griggs's letter is in the minutes of March 10, 1913.

6. Letter, February 18, 1913, in minutes of February 25, 1913, St. Paul & Tacoma Lumber Co. Records.

7. Letter, February 20, 1913, in minutes of March 5, 1913, St. Paul & Tacoma Lumber Co. Records.

8. Griggs to Jones, in minutes of March 5, 1913, St. Paul & Tacoma Lumber Co. Records.

9. Minutes, February 10 and March 10, 1913, St. Paul & Tacoma Lumber Co. Records.

10. Minutes, March 10, 1913, St. Paul & Tacoma Lumber Co. Records.

11. Minutes, January 14, 1914, St. Paul & Tacoma Lumber Co. Records.

12. Minutes, annual meeting, January 31, 1914, St. Paul & Tacoma Lumber Co. Records.

13. The best account of the Howarths is in an autobiographic letter written by William Howarth to Richard Heape of Rochdale, England, on May 25, 1922, in which he discusses his and Leonard's activities in America. See also William Prosser, A History of the Puget Sound Country (New York: Lewis Publishing Co., 1903), pp. 567–68; Seattle Post-Intelligencer, February 13, 1937 (obituary), West Coast Lumberman, March 1930, p. 33 (obituary).

14. Minutes, January 31, 1914, St. Paul & Tacoma Lumber Co. Records.

15. Minutes, February 27, 1914, St. Paul & Tacoma Lumber Co. Records.

16. Hidy et al., Timber and Men, pp. 273–77.

St. Paul & Tacoma at War

1. Norman Clark, Mill Town (Seattle: University of Washington Press, 1970), pp. 131–32.

2. Robert E. Ficken, Lumber and Politics: The Career of Mark E.

Reed (Seattle: University of Washington Press, 1979), p. 134; William G. Reed with Elwood R. Maunder, *Four Generations of Management: The Simpson-Reed Story* (Santa Cruz: Forest History Society, 1977), p. 49; Russell Earley, "St. Paul & Tacoma Lumber Company," unnumbered typescript, St. Paul & Tacoma Lumber Co. Records, University of Washington Libraries.

3. Ralph Hidy, Frank Hill, and Allan Nevins, *Timber and Men: The Weyerhaeuser Story* (New York: Macmillan Co., 1963), p. 334; *American Lumberman,* March 31, 1917, p. 24; April 21, 1917, p. 35; *The Timberman,* August 1917.

4. Ellis Lucia, *Head Rig: The Story of the West Coast Lumber Industry* (Portland, Ore.: Overland West Press, 1965), pp. 87–88; Hidy et al., *Timber and Men,* pp. 335–37.

5. Earley, "St. Paul & Tacoma Lumber Company.

6. *American Lumberman,* April 28, 1917, p. 24.

7. C. Bradford Mitchell, *Every Kind of Shipwork* (New York: Todd Shipyards Corp., 1981), pp. 34–43; Hidy et al., *Timber and Men,* p. 334; Lucia, *Head Rig,* p. 88.

8. Minutes, October 17, 1917 and February 4, 1919, St. Paul & Tacoma Lumber Co. Records, University of Washington Libraries.

9. Vernon H. Jensen, *Lumber and Labor* (New York: Farrar & Rinehart, 1945), pp. 129–37; Hidy et al., *Timber and Men,* pp. 335–43.

10. Ficken, *Lumber and Politics,* p. 35; Hidy et al., *Timber and Men,* p. 339.

11. Hidy et al., *Timber and Men,* p. 343; Lucia, *Head Rig,* p. 94.

12. Vernon H. Jensen, "Labor Relations in the Douglas Fir Lumber Industry," Ph.D. dissertation, University of California, 1939, p. 63; Jensen, *Lumber and Labor,* p. 129; C. R. Howd, *Industrial Relations in the West Coast Lumber Industry,* bulletin 349, U.S. Dept. of Labor, Bureau of Labor Statistics (Washington, D.C.: U.S. Government Printing Office, 1924), p. 77.

13. Charles M. Gates, *The First Century at the University of Washington* (Seattle: University of Washington Press, 1961), pp. 151–53; Hidy et al., *Timber and Men,* p. 345; Ficken, *Lumber and Politics,* pp. 36–37.

14. Jensen, *Lumber and Labor,* p. 132; Hidy et al., *Timber and Men,* p. 345.

15. Jensen, *Lumber and Labor,* p. 108; Hidy et al., *Timber and Men,* pp. 347–48; Ficken, *Lumber and Politics,* pp. 36–39; Ruby El Hult, *The Untamed Olympics* (Portland, Ore.: Binfords and Mort, c. 1971), pp. 198, 199, 204; Lucia, *Head Rig,* p. 93.

16. Lucia, *Head Rig,* p. 94.

17. Ibid., pp. 95–97.

State of the Art: Mill C

1. *Tacoma Ledger,* January 17, 1917.

2. *Tacoma Ledger,* May 3, 1919.

3. This chapter is based almost entirely on an advertising supplement commissioned by the St. Paul & Tacoma in *American Lumberman,* May 21, 1921.

Tokyo, New York, Sunnyside, and Way Points

1. Ellis Lucia, *Head Rig: The Story of the West Coast Lumber Industry* (Portland, Ore.: Overland West Press, 1965), pp. 147–48.

2. William G. Reed with Elwood Maunder, *Four Generations of*

Management: The Simpson-Reed Story (Santa Cruz: Forest History Society, 1977), pp. 72–73; Robert E. Ficken, *Lumber and Politics* (Seattle: University of Washington Press, 1979), p. 133.

3. Ellis Lucia, *Head Rig,* p. 148.

4. *West Coast Lumberman,* May 1, 1924.

5. Everett Griggs to Mark Reed, September 22, 1927. St. Paul & Tacoma Lumber Co. Records, University of Washington Libraries.

6. Ficken, *Lumber and Politics,* p. 140.

7. Russell Earley, "St. Paul & Tacoma Lumber Company," unnumbered typescript, St. Paul & Tacoma Lumber Co. Records, University of Washington Libraries; interview with Corydon Wagner, Jr.

8. Minutes, August 27, 1926, St. Paul & Tacoma Lumber Co. Records, University of Washington Libraries.

9. Earley, "St. Paul & Tacoma Lumber Company"; *American Lumberman,* May 21, 1921.

10. Earley, "St. Paul & Tacoma Lumber Company."

Keeping Up with the Joneses

1. *Tacoma Ledger,* November 29, 1922; *Grays Harbor Washingtonian,* November 29, 1922.

2. W. C. Pettit, "History of the Northwestern Lumber Company," in *Washington State Industries and Resources: Lumber Mill Histories,* typescript, Hoquiam Public Library; *Encyclopedia of Biography,* 12:346.

3. Egbert S. Oliver, "Sawmilling on Grays Harbor in the Twenties: A Personal Reminiscence," *Pacific Northwest Quarterly* 69 (January 1978): 1–18.

4. *Grays Harbor Washingtonian,* May 23, 24, 1918.

5. Edward H. Dodd, "History of the College of Puget Sound," unpublished manuscript in the University of Puget Sound Library, p. 251.

6. Ibid., pp. 250–51.

7. *Tacoma Ledger,* May 22, 1923.

8. Pettit, "History of the Northwestern Lumber Company."

9. *Tacoma Sunday Ledger* and *News-Tribune,* February 4, 1925.

10. *Tacoma Ledger,* April 25, 1931.

Hemlock, the Wonderful Weed

1. Williams Haynes, *Cellulose: The Chemical That Grows* (New York, 1953), pp. 17–26.

2. Russell Earley, "St. Paul & Tacoma Lumber Company," unnumbered typescript, St. Paul & Tacoma Lumber Co. Records, University of Washington Libraries (chapter entitled "Reinventory of Timber Resources").

3. Ibid.; minutes, September 7, 1927, St. Paul & Tacoma Lumber Co. Records, University of Washington Libraries.

4. Minutes, October 2, 1927, St. Paul & Tacoma Lumber Co. Records.

5. Earley, "St. Paul & Tacoma Lumber Company" (chapter entitled "Union Bag and Paper").

Years of Change

1. Robert E. Ficken, *Lumber and Politics: The Career of Mark E. Reed* (Seattle: University of Washington Press, 1979), p. 154; *Mason County Journal,* June 21, 1927.

2. Ficken, *Lumber and Politics,* pp. 194–95; William G. Reed with Elwood Maunder, *Four Generations of Management: The Simp-*

son-Reed Story (Santa Cruz: Forest History Society, 1977), pp. 83–84; Minot Davis to Mark Reed, March 3 and May 27, 1931 (quoted in Ficken, *Lumber and Politics*); Simpson Logging Company Papers; Rodney C. Loehr, ed., *Forests for the Future: The Story of David T. Mason, 1907–50.* (St. Paul: Minnesota Historical Society, 1952).

3. Ralph W. Hidy, Frank E. Hill, and Allan Nevins, *Timber and Men: The Weyerhaeuser Story* (New York: Macmillan Co., 1963), pp. 437–40; Ellis Lucia, *Head Rig: The Story of the West Coast Lumber Industry* (Portland, Ore.: Overland West Press, 1965), pp. 17–73 passim.

4. *Tacoma News-Tribune*, August 18, 19, 20, 21, 1933; *Tacoma Ledger*, August 19, 22, 1933.

5. Washington State Labor News, February 23, 1934; Vernon H. Jensen, *Lumber and Labor* (New York: Farrar & Rinehart, 1945), pp. 159–60.

6. *Timberworker*, September 7, 1936; *Washington State Labor News*, March 29, 1935; *West Coast Lumberman*, June 1935, p. 26; Jensen, *Lumber and Labor*, pp. 164–65; Hidy et al., *Timber and Men*, pp. 424–25.

7. Hidy et al., *Timber and Men*, p. 426; Jensen, *Lumber and Labor*, p. 167.

8. Thirty-third Semi-annual Report for the Board of Directors, May 20–22, 1935, pp. 1–4, St. Paul & Tacoma Lumber Co. Records, University of Washington Libraries; *Tacoma Times*, May 6, 10, 13, 21 and June 4, 1935; *Tacoma News-Tribune*, May 6, 7, 8, 12, 21, 22, 1935; *Four-L News Letter*, June 4, 1935.

9. *Tacoma Times*, June 7, 1935.

10. *Seattle Times*, June 8, 10, 1935; *Tacoma News-Tribune*, June 9, 10, 11, 1935.

11. *Tacoma Times*, June 21–23, 1935; *Tacoma News-Tribune*, June 21–23, 1935.

12. *Tacoma News-Tribune*, June 24, 1935; *Seattle Post-Intelligencer*, June 24, 1935.

13. *Seattle Post-Intelligencer*, June 27, 1935; *Tacoma Labor Advocate*, June 30, 1935.

14. Jensen, *Lumber and Labor*, p. 180.

15. *Tacoma News-Tribune*, July 11, 13, 1935; *Seattle Post-Intelligencer*, July 12, 13, 1935; *Tacoma Labor Advocate*, July 13, 1935; *Tacoma Times*, July 13, 1935.

16. Vernon H. Jensen, "Labor Relations in the Douglas Fir Industry," Ph.D. dissertation, University of California, 1939, p. 261.

17. Russell Earley, "St. Paul & Tacoma Lumber Company," unnumbered typescript, St. Paul & Tacoma Lumber Co. Records, University of Washington Libraries (chapter on "American Lumber Standards Handbook").

18. Hidy et al., *Timber and Men*, p. 446.

19. Ibid., p. 447; Earley, "St. Paul & Tacoma Lumber Company" ("Conspiracy Trials of Lumber Industry"); Lucia, *Head Rig*, pp. 199–207.

20. West Coast Lumbermen's Association news release, April 1941.

The Renewable Resources

1. *American Lumber Industry*, Proceedings of the National Lumber Manufacturers' Association, May 1912 (Cincinnati, Ohio).

2. Russell Earley, "St. Paul & Tacoma Lumber Company," unnumbered typescript, St. Paul & Tacoma Lumber Co. Records, University of Washington Libraries (chapter entitled "Reinventory of Timber Resources and Reappraisal of Timberlands").

3. Ellis Lucia, *Head Rig: The Story of the West Coast Lumber Industry* (Portland, Ore.: Overland West Press, 1965), pp. 191–98.

Only Change Goes On Forever . . .

1. Joseph Davis, *The Coal Fields of Pierce County,* Washington Geological Survey Bulletin 10 (Olympia, Washington, 1914).

2. E. V. Smalley, comp., "Book of Reference for the Use of the Directors and Officers" (St. Paul, Minn., 1883), Northern Pacific Archives. (Entries for February 16, August 16, 1876; March 21, May 9, 1877; January 3, February 19, 1878.)

3. Minutes, May 31, 1890, St. Paul & Tacoma Lumber Co. Records, University of Washington Libraries.

4. "The Wilkeson Products Company, Successors to the Wilkeson Coal and Coke Company," reports of state mining inspectors, 1888–1943, on the Wilkeson Coal and Coke properties (7-page typescript), St. Paul & Tacoma Lumber Co. Records, University of Washington Libraries.

5. Interview with Chauncey L. Griggs.

6. Minutes of Special Meeting of Directors, Wilkeson Products Company, November 24, 1944.

7. "Coast Mines Set Wilkeson to Growing," *Tacoma Sunday Ledger,* September 4, 1966; Davis, *Coal Fields of Pierce County.*

8. Interview with Chauncey L. Griggs.

The Final Chapter

1. The outline of events and debates leading to the sale of the St. Paul & Tacoma Lumber Company to St. Regis is to be found in the minutes of the board (St. Paul & Tacoma Lumber Co. Records, University of Washington Libraries). This skeleton has been fleshed out with detail from interviews with surviving figures in the drama, among them: Corydon Wagner, Jr. (now deceased), Corydon Wagner III, Henry Hewitt, Jr., Chauncey Griggs, and Anson Moody. I also had access to some material from the personal files of Wagner, Griggs, and Hewitt that are not part of the St. Paul & Tacoma records.

2. Griggs to Wagner, March 1950, copy of the memo with the minutes of the annual meeting, St. Paul & Tacoma Lumber Co. Records.

3. Wagner to Griggs, March 1950, copy with minutes of annual meeting.

4. Wagner to St. Paul & Tacoma Board, March 1950, copy with minutes of annual meeting.

5. Minutes of the annual meeting, March 1951, St. Paul & Tacoma Lumber Co. Records.

6. Robert M. Court, *The Plywood Age* (Portland, Ore., 1955), pp. 44–45.

7. Minutes, August 1946, and Russell Earley, "St. Paul & Tacoma Lumber Company," unnumbered typescript, St. Paul & Tacoma Lumber Co. Records, University of Washington Libraries.

8. Eleanor Amigo and Mark Neuffer, with Elwood R. Maunder, *Beyond the Adirondacks: The Story of St. Regis Paper Company* (Westport, Conn.: Greenwood Press, c. 1980), pp. 85, 90–91.

9. Interviews with Anson Moody and Chauncey Griggs.

10. Interview with Chauncey Griggs.

11. Interview with Henry Hewitt, Jr.

12. Interviews with Hewitt and Corydon Wagner III. (Hewitt and Wagner retain reservations about Lindberg; Chauncey Griggs still admires him.)

13. Interviews with Wagner and Hewitt.

14. Interview with Anson Moody.

15. Interview with Chauncey Griggs.

16. Interview with Henry Hewitt, Jr.; personal file of Hewitt notes summarizes meetings and conversations during the period in question.

17. Interview with Chauncey Griggs.

18. Interview with Griggs and Anson Moody.
19. Interview with Chauncey Griggs.
20. Interview with Anson Moody.
21. C. W. Griggs Investment Company minutes.
22. Minutes, January 1956, St. Paul & Tacoma Lumber Co. Records.
23. Minutes, February 25, 1957, St. Paul & Tacoma Lumber Co. Records.

Loggers bid farewell to a stand of old growth (from West Coast Lumberman)

Index